Current Topics in Microbiology and Immunology

Volume 434

The reviews series *Current Topics in Microbiology and Immunology* publishes cutting-edge syntheses of the latest advances in molecular immunology, medical microbiology, virology and biotechnology. Each volume of the series highlights a selected timely topic, is curated by a dedicated expert in the respective field, and contains a wealth of information on the featured subject by combining fundamental knowledge with latest research results in a unique manner.

2020 Impact Factor: 4.291, 5-Year Impact Factor: 5.110

2020 Eigenfaktor Score: 0.00667, Article Influence Score: 1.480

2020 Cite Score: 7.7, h5-Index: 38

More information about this series at https://link.springer.com/bookseries/82

Takashi Nagasawa
Editor

Bone Marrow Niche

Microenvironments Critical for Immune Cell Development

Responsible Series Editor: Shizuo Akira

Springer

Editor
Takashi Nagasawa
Graduate School of Frontier Biosciences
and Graduate School of Medicine
Osaka University
Osaka, Japan

ISSN 0070-217X ISSN 2196-9965 (electronic)
Current Topics in Microbiology and Immunology
ISBN 978-3-030-86015-8 ISBN 978-3-030-86016-5 (eBook)
https://doi.org/10.1007/978-3-030-86016-5

This Springer imprint is published by the registered company Springer Nature Switzerland AG
The registered company address is: Gewerbestrasse 11, 6330 Cham, Switzerland

Preface

Almost all blood cells, including immune cells, are generated from hematopoietic stem cells (HSCs) in bone marrow and maintained in various lymphoid organs. However, HSCs and immune cells cannot survive on its own and thus are in contact with and require the special microenvironment termed niches for their maintenance, proliferation, and differentiation in the bone marrow and lymphoid organs. In particular, the identity of HSC niches in bone marrow had been a subject of longstanding debate. It has been thought that HSC niches comprise various types of support cells that provide critical signals, including cytokines and extracellular matrix for HSC regulation.

Bone lining osteoblasts are the first support cell population reported to influence HSCs; however, recent studies have clearly demonstrated that the population of bone marrow-specific stromal cells, termed CXC chemokine ligand 12 (CXCL12)-abundant reticular (CAR) cells, which overlap strongly with leptin receptor-expressing (LepR$^+$) cells, are the major cellular component of niches for HSCs and immune cells. CAR/LepR$^+$ cells are skeletal stem cells that are scattered throughout marrow cavity and have potential to give rise to osteoblasts and adipocytes located in the endosteum and marrow cavity, respectively, throughout life. All CAR/LepR$^+$ cells have long processes and express much higher levels of HSC niche factors, including cytokines CXCL12 and stem cell factor (SCF) and transcription factors (Foxc1 and Ebf3) than other cell population. About 62% of CAR/LepR$^+$ cells express interleukin (IL)-7 essential for lymphopoiesis and two subclusters within the population were isolated and termed Adipo-CAR and Osteo-CAR cells. The Osteo-CAR cells comprise about 7% of CAR/LepR$^+$ cells and express higher levels of Osterix and lower levels of LepR than Adipo-CAR cells. The candidate developmental origin of skeletal stem and progenitor cells, including CAR/LepR$^+$ cells, have been extensively investigated. In addition, various types of cells, including adipocytes, sinusoidal endothelial cells, Myh11$^+$NG2$^+$ periarteriolar cells, PDGFRα^+Sca-1$^+$CD45$^-$Ter119$^-$ cells termed PαS cells, megakaryocytes, and αSMA$^+$ macrophages, have been reported to be candidate HSC niche cells. Of note, recent studies revealed that the features and functions of HSC niche cells change in inflammation, infection and hematological malignancies.

Furthermore, in extramedullary lymphoid tissues, population of lymphoid tissue stromal cells, including T zone reticular cells (TRC), CXCL12-expressing reticular cells (CRC), follicular dendritic cells (FDC), marginal reticular cells (MRC), and medullary cord fibroblastic reticular cells (MedRC) in lymph nodes and $WT1^{+}$ stromal cells in omentum, have been identified as candidate cellular niches for immune cells. Interestingly, some of these cells share features and markers for HSC niche cells in the bone marrow. In this book, we focus on cellular niches for HSCs and/or immune cells in the bone marrow and extramedullary lymphoid tissues and summarize recent advances in research, comparing their features and functions that would contribute to the emergence of new concepts of lymphohematopoietic microenvironments.

Osaka, Japan Takashi Nagasawa

Contents

Skeletal Stem Cells as the Developmental Origin of Cellular Niches for Hematopoietic Stem and Progenitor Cells 1
Thomas H. Ambrosi and Charles K. F. Chan

Cellular Niches for Hematopoietic Stem Cells and Lympho-Hematopoiesis in Bone Marrow During Homeostasis and Blood Cancers 33
Yoshiki Omatsu, Kei Higaki, and Takashi Nagasawa

Remodeling of the Bone Marrow Stromal Microenvironment During Pathogenic Infections 55
YeVin Mun, Serena Fazio, and César Nombela Arrieta

The Roles of IL-7 and IL-15 in Niches for Lymphocyte Progenitors and Immune Cells in Lymphoid Organs 83
Koichi Ikuta, Takahiro Hara, Shinya Abe, Takuma Asahi, Daichi Takami, and Guangwei Cui

Lymph Node Stromal Cells: Diverse Meshwork Structures Weave Functionally Subdivided Niches 103
Arata Takeuchi, Madoka Ozawa, Guangwei Cui, Koichi Ikuta, and Tomoya Katakai

Immune Niche Within the Peritoneal Cavity 123
Yasutaka Okabe

Skeletal Stem Cells as the Developmental Origin of Cellular Niches for Hematopoietic Stem and Progenitor Cells

Thomas H. Ambrosi and Charles K. F. Chan

Abstract The skeletal system is a highly complex network of mesenchymal, hematopoietic, and vasculogenic stem cell lineages that coordinate the development and maintenance of defined microenvironments, so-called niches. Technological advancements in recent years have allowed for the dissection of crucial cell types as well as their autocrine and paracrine signals that regulate these niches during development, homeostasis, regeneration, and disease. Ingress of blood vessels and bone marrow hematopoiesis are initiated by skeletal stem cells (SSCs) and their more committed downstream lineage cell types that direct shape and form of skeletal elements. In this chapter, we focus on the role of SSCs as the developmental origin of niches for hematopoietic stem and progenitor cells. We discuss latest updates in the definition of SSCs, cellular processes establishing and maintaining niches, as well as alterations of stem cell microenvironments promoting malignancies. We conclude with an outlook on future studies that could take advantage of SSC-niche engineering as a basis for the development of new therapeutic tools to not only treat bone-related diseases but also maladies stemming from derailed niche dynamics altering hematopoietic output.

1 Introduction

The skeleton is maintained by constant bone remodeling facilitated through a balance of bone formation and resorption processes. Bones are made up from many different cell types including osteocytes, chondrocytes, endothelial cells, hematopoietic and immune cells, and nerve cells as well as their stem and progenitor populations. Each cell type resides in specific microenvironments and their interactions orchestrate niche function. Niche cells are able to sense signals from their surroundings that

T. H. Ambrosi · C. K. F. Chan (✉)
Institute for Stem Cell Biology and Regenerative Medicine, Stanford University School of Medicine, Stanford, CA, USA
e-mail: chazchan@stanford.edu

Department of Surgery, Stanford University School of Medicine, Stanford, CA 94305, USA

T. Nagasawa (ed.), *Bone Marrow Niche*, Current Topics in Microbiology and Immunology 434, https://doi.org/10.1007/978-3-030-86016-5_1

allows them to adjust crosstalk with each other thereby directing activity in a localized manner (Pinho and Frenette 2019).

Stem cells are particularly important for the replenishment of mature cell types that are required for the growth, maintenance, and regeneration of adult organ tissues. The high dynamism of bone tissue, which turns over completely each year in children and continues in adults at progressively slower rates, is a consequence of particularly pronounced stem cell activity (Demontiero et al. 2012). Skeletal stem cells (SSCs) are the source of bone, cartilage, and stromal cells which are crucial for shaping skeletal structures (Ambrosi et al. 2019). They also create the bone marrow cavity which is the main site of hematopoiesis during postnatal life and in which the majority of hematopoietic stem cells (HSCs) resides (Chan et al. 2009, 2015). HSCs give rise to the progenitor cells of the hematopoietic and immune system including all cell types encompassing erythrocytes, platelets, and the myeloid and lymphoid arms such as neutrophils, macrophages, natural killer cells, and T/B cells (Weissman and Shizuru 2008). By the sheer number of different cell types, it becomes clear that a tight coordination of cellular output is necessary. The most extensively studied bone marrow microenvironments are those related to providing signals for quiescence, replication, and differentiation of HSCs. The hematopoietic stem cell niche is largely shaped and controlled by SSC-derived lineage cells. Thus, SSC and HSC lineage interactions control anabolic and catabolic processes of bones through intertwined actions of skeletal tissue forming SSC cell populations with hematopoietic cell-derived bone-resorbing osteoclasts and simultaneous hematopoiesis. This underscores the multiple vital aspects of a proper control of niche functions for preserving health.

Aging and disease can disrupt niche dynamics that can lead to bone loss and have broader implications for systemic health (Beerman et al. 2010; Jaiswal and Ebert 2019). The spectrum of diseases spans from osteoporosis and nonunions after fracture, to autoimmune malignancies such as rheumatoid arthritis and inflammatory bowel disease, and even cancer. The combined loss of osteochondrogenic differentiation potential and promotion of osteoclasts maturation and activity drives exacerbated bone loss with age (Ambrosi et al. 2021a). Elucidating the signaling factors involved and identifying new ways to engineer niche function to reinstate youthful states might allow for novel therapeutic approaches counteracting maladies stemming from a dysregulated microenvironment. The clinical potential of HSCs has long been realized and has helped treating cancers, neural disorders, immunodeficiency, or genetic disorders (Weissman and Shizuru 2008). The common approach to treat patients consists of the ablation of a recipient's immune system by radiation or chemotherapy with subsequent transplantation of healthy hematopoietic stem and progenitor cells (HSPCs) to replenish a derailed blood system. This is possible since HSCs possess the ability to home to their niches in the bone marrow by traveling through the circulation. In contrast, SSCs and their downstream progenitors or populations enriched for these cells, also commonly called "mesenchymal stromal cells" ("MSCs") that provide those homing signals lack effective re-engraftment at endogenous sites if injected systemically (Bianco and Robey 2015). This is important because most clinical efforts have been focused on expanding and/or manipulating mesenchymal cell populations ex vivo before transplanting them into a sick patient. As a result, treatment options involving mesenchymal cell types remain limited. A more effective alternative to skeletal stem cell transplantation would be to focus on signals that regulate resident SSC lineage cells to direct these cell types to expand, differentiate, and regenerate tissues. Potential approaches are numerous and could involve biomaterials, engineered immune cells for targeted delivery of factors, or nanobodies and small molecules. In order to therapeutically leverage the regenerative power of stem cells, we need to better understand SSC characteristics, the cellular localization and composition of skeletal niche microenvironments, and systemic factors that control niche dynamics.

2 Skeletal Stem Cells

2.1 Definition and Nomenclature

Adult tissues contain multipotent stem cells that facilitate organ maturation, maintenance, and regeneration. Studying stem cell biology is fundamental for the development of new therapies as it gives insight into development, cellular differentiation, and tissue function. Ernst Haeckel was among the first scientists to suggest that all living cells undergo an evolutionary sequence originating from a unicellular ancestor (Dose 1981). The study of the origin of the blood system, however, paved the way to our understanding of the term stem cell. Artur Pappenheim first used it to designate the cell at the apex of hematopoiesis before many others came to support his view in the early years of the twentieth century (Pappenheim 1896). It took several decades more until Till and McCulloch were able to experimentally address and prove that a single rare bone marrow cell gives rise to multiple blood lineages (Till and McCulloch 1961; Becker et al. 1963). Technological advances allowing the prospective isolation of phenotypically defined cell populations via flow cytometry created the basis for the identification of cell populations enriched for HSCs as well as many of the downstream cell types needed for a functioning blood and immune system (Spangrude et al. 1988; Laurenti and Göttgens 2018). The current definition of a single HSC with bona fide stem cell characteristics entails the ability to self-renew, to give rise to all cell populations of the hematopoietic lineage tree, and to be the source of long-term reconstitution of the hematopoietic system after hematopoietic ablation in serial transplants in vivo (Morrison et al. 1997; Weissman and Shizuru 2008).

Bone marrow is the primary site of postnatal hematopoiesis. The non-hematopoietic stroma provides structural and biochemical cues that are crucial for HSC-mediated processes. As a result, much effort has been directed towards delineating the cellular origin of skeletal tissue. Like many cell types, bone progenitors also have the ability to adhere to plastic upon isolation and plating to culture vessels. Discoveries by Friedenstein and others showed that these cells possess fibroblast colony forming unit (CFU-F) ability and multi-differentiation capacity into at least bone, cartilage, and adipocyte lineages (Friedenstein et al. 1966; Castro-Malaspina et al. 1980; Kuznetsov et al. 1997). The concept of self-renewing, clonogenic, multipotent bone marrow stromal cells (BMSCs) was definitively established and incorporated with general guidelines for their definition by International Society of Cellular Therapies (ISCT) shortly after (Caplan 1991; Pittenger et al. 1999; Dominici et al. 2006). In an allusion to HSCs, the term "Mesenchymal Stem/Stromal Cell" ("MSC") is often used to describe this analogous bone-forming population. However, in contrast to HSCs which were functionally defined by in vivo self-renewal and serial transplantation under limiting dilution conditions, the criteria to define "MSCs" have long been based on in vitro assays known to alter endogenous cell properties including differentiation capacity and cell surface marker expression. Under highly artificial and super physiological in vitro stimuli, "MSCs" have been shown to not only differentiate into bone, cartilage, and fat but also tendon, vascular, muscle, and nerve tissue (Crisan et al. 2008; Via et al. 2012). Since similar experimental readouts have been made for plastic-adherent cells from various connective tissues, e.g., muscle and adipose tissue, it has been generally assumed that these cells can interchangeably be used for tissue-specific regenerative approaches. A commonality between "MSCs" of different tissues is their perivascular localization (Caplan 2008). Strikingly, stringent in vivo experiments have demonstrated that this ubiquitous pericyte differs in its transcriptional signature and differentiation capacity depending on anatomical sources (Sacchetti et al. 2016). Thus, infiltration of blood vessels during organ development might lead to simultaneous migration of mural cells into the connective tissues which become permanently primed to tissue-specific lineages due to local morphogen exposure (Urist 1970; Pearson et al. 1986; James et al. 2010; Corselli et al. 2012; Salazar et al. 2016). This might explain why harnessing the potential clinical power of "MSCs" has not been fully leveraged. Taken together, those observations have resulted in frequent contradictory findings leading the field to eventually decide to refrain from the term "MSC" and assign tissue-specific nomenclature (Bianco and Robey 2015; Caplan 2017; Sipp et al. 2018; Ambrosi et al. 2019), i.e., skeletal stem cell, muscle stem cell, adipose stem cell, etc.

To test the in vivo properties of BMSCs in the same manner as HSCs, new assays needed to be developed. In vivo transplantation experiments with and without previous in vitro expansion now show that BMSCs are restricted to skeletogenic lineages and that more rigorously defined skeletal stem and progenitor cell types can self-renew and generate bone, cartilage, and stroma in vivo on the single cell level (Sacchetti et al. 2007; Tormin et al. 2011; Chan et al. 2013). Recent work also has identified a bona fide mouse and human skeletal stem cell (SSC) which gives rise to bone, cartilage, and stroma (Chan et al. 2015, 2018). In agreement with the findings of other studies, an SSC is now defined as a bone-resident stem cell committed to skeletogenesis and able to recapitulate bone organogenesis in vivo (Sacchetti et al. 2007; Méndez-Ferrer et al. 2010; Bianco 2014). A single SSC, both self-renews and initiates clonal propagation with subsequent differentiation to all skeletogenic lineages, and shows evidence of long-term renewal either under serial passaging or serial transplantation conditions. Ossicles formed by SSCs upon in vivo transplantation, either subcutaneously or under the renal capsule, represent chimeras consisting of SSC-derived bone and bone marrow stroma as well as host-derived hematopoietic tissue and blood vessels (Gulati et al. 2018). Therefore, SSCs are responsible for maintaining skeletal physiology, including growth, turnover, regeneration, and regulation of the hematopoietic microenvironment. Nonetheless, these assays are subject to the same caveats used to determine stem cell characteristics of HSCs in bone marrow transplant in that they are founded on assays involving tissue perturbations. In the current approach, the SSCs have to be removed from their endogenous environment to be tested. An ideal method would be to test SSC function in situ, using genetic lineage tracing which would enable fate mapping of skeletal cell populations labeled by a single marker. However, genetic tracing specific enough to follow single SSCs in their endogenous setting has not been convincingly achieved (Zhou et al. 2014; Mizuhashi et al. 2018). Reasons for this include the complex bone composition and the lack of truly SSC-specific markers. Future technological advancements will be needed to refine these common approaches for characterizing pure SSC populations.

2.2 *Origin*

SSC multipotency is limited to skeletal lineage tissues such as bone, cartilage, and hematopoiesis-supporting stroma. With the exception of germline stem cells, postnatal vertebrates do not possess a pluripotent stem cell reservoir. Every organism begins as a totipotent zygote, the one-cell-stage after fertilization of the egg, and continuously gives rise to more specialized cells with restricted differentiation capacities (Evans and Kaufman 1981). Through the embryonic process of gastrulation, three distinct germ layers emerge; the ectoderm which forms the outer body covering including skin and neural cells, the mesoderm which develops pivotal cell types of most organs such as muscle and blood cells, and the endoderm which gives rise to the inner lining of tissues associated with the digestive and respiratory systems (Murry and Keller 2008). The developmental origin of skeletogenic cell types varies between anatomical location of bones but yields the same mature bone-forming osteoblasts and osteocytes. The majority of the skeleton derives from two specifications of the mesoderm, the paraxial (axial skeleton, e.g., vertebrae, ribs) or lateral plate (appendicular skeleton, e.g., limbs) mesoderm, whereas facial and skull bones mainly originate from ectoderm-derived neural crest (Olsen et al. 2000). Despite having the potential to generate all skeletal lineages through a sequence of fate choices, the developmental time point determines the proportion of SSC-derived specialized cell types that are produced. Additionally, there are regional-specific differences in SSCs and their contribution to mature cell populations of the bone (Ambrosi et al. 2021b; Matthews et al. 2021; Sivaraj et al. 2021; Shu et al. 2021). Some bones of the skull form through intramembranous ossification in which ossification proceeds from a fibrous intermediate. However, the predominant bone-forming process in the paraxial skeleton involves endochondral ossification, which proceeds from a cartilage template containing a series of proliferative, pre-hypertrophic, and hypertrophic chondrocytes that is subsequently replaced with bone-matrix producing osteogenic cells in the calcification zone. SOX9 (SRY-Box Transcription Factor 9)-expressing SSCs line the cartilage anlage as part of the perichondrium acting as a replenishing source for the primary ossification center. Blood vessel infiltration not only drives ossification but also the differentiation of SSCs into stromal lineages. Importantly, bone marrow hematopoiesis does not occur before the emergence of marrow stroma shortly before birth and remains minimal in bones formed through intramembranous ossification (Chan et al. 2009; Rowe et al. 2016). Perinatally, vascular invasion into the distal ends of long bones creates secondary ossification centers creating the epiphyseal endplates containing spongy bone with no medullary cavity and articular cartilage. Interestingly, it was recently proposed that at least a fraction of BMSCs that contribute to osteogenic and stromal lineages during establishment of bone marrow hematopoiesis arise via endothelial-hematopoietic transition (Kenswil et al. 2021). All in all, the identity of a universal bona fide SSC population as a source for all subregions of the skeletal compartment with its different tissues made of the skeletogenic lineages remains controversial.

2.3 *Identity*

The constant growth and remodeling of the skeleton late into adulthood with its diverse anatomical subregions and dynamically changing structural composition creates a great demand for locally controlled replenishment with skeletogenic cell types. In line with this diversity, detailed studies on the identity and anatomical localization of skeletal stem and progenitor cells have found BMSCs to be enriched in vascular, periosteal, growth plate, and articular cartilage regions. Recent single cell RNA-sequencing data of non-hematopoietic bone marrow fractions further supports the existence of multiple SSC subsets but also demonstrates limitations of available SSC markers in delineating homogenous SSC populations (Baryawno et al. 2019; Tikhonova et al. 2019; Wolock et al. 2019). From a historical standpoint and as described above due to the nature of mostly defining SSCs based on in vitro assays upon plastic adherence, the minimal SSC criteria are considered CFU-F ability and tri-lineage differentiation capacity into osteogenic, chondrogenic, stromal, and possibly adipogenic fates. Aside from testing those properties in vitro, they are usually also only performed on bulk cell populations. Thus, many reported putative SSC populations are very heterogeneous reflecting the combined properties of multiple types of lineage-restricted progenitor cells (Fig. 1).

Recent technical advances including genetic lineage tracing in rodents and the prospective isolation of cell populations via fluorescence-activated cell sorting (FACS) based on antibody staining against cell surface marker configurations have substantially clarified the true lineage potential of SSCs. These techniques now provide evidence that growth plate SSCs in mice and humans contribute to bone, cartilage, and reticular stroma, but not bone marrow adipose tissue (BMAT) (Chan et al. 2015, 2018; Worthley et al. 2015; Mizuhashi et al. 2018; Ambrosi et al. 2021b). FACS has been a particularly useful tool to target defined cell populations in human

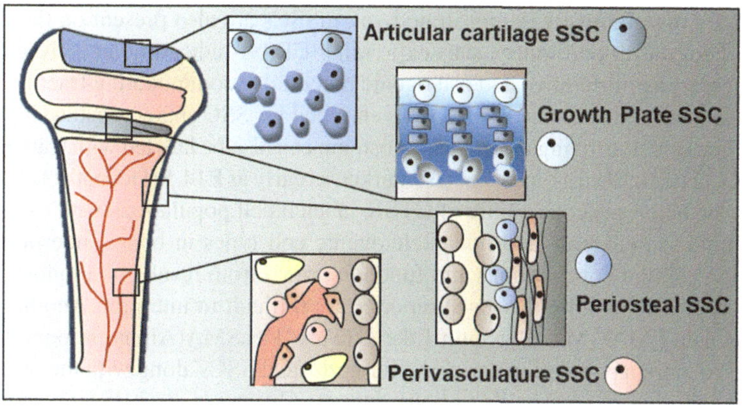

Fig. 1 Multiple Skeletal Stem Cell (SSC) Subtypes. SSCs of long bones are found in specific anatomical localizations. Their microenvironment determines lineage contributions and functions

tissue. Highly osteochondrogenic SSCs in the developing bone as well as in postnatal growth plates and fracture tissue were identified in the non-hematopoietic, nonendothelial fraction as $CD146^{-}PDPN(Podoplanin)^{+}CD164^{+}CD73^{+}$ (hSSC), or $6C3^{-}THY1(CD90)^{-}CD105^{-}CD51^{+}CD200^{+}$ (mSSC) in humans and mice, respectively (Chan et al. 2015, 2018; Ambrosi et al. 2020). Articular cartilage also harbors a clonogenic subset of these SSCs (Murphy et al. 2020). The best genetic markers employed to target mouse growth plate-specific SSCs to date are Gremlin1 (Grem1) and Parathyroid hormone-related protein (PTHrP) (Worthley et al. 2015; Mizuhashi et al. 2018). Fate mapping experiments using the inducible PTHrP–CreER mice crossed with a tomato reporter line revealed only limited contribution of those growth plate SSCs to trabecular bone and bone marrow stroma. This could be a consequence of incomplete recombination, part of the limitations when interrogating SSC biology with genetic mouse models (discussed elsewhere (Ambrosi et al. 2019)), but also support the view that SSC tissue contributions are indeed regionally restricted. Two recent studies using dual labeling approaches indeed found that metaphyseal and diaphyseal BMSCs differentially contribute to skeletogenic tissues in a spatiotemporal manner (Sivaraj et al. 2021; Shu et al. 2021). Lineage tracing in postnatal long bones suggested that growth plate BMSCs with chondrocyte phenotype labeled by Aggrecan (Acan) expression facilitate bone lengthening while BMSCs labeled by Leptin Receptor (LepR) contribute to bone thickening.

Similarly, functional and anatomical restrictions have been found in the study of periosteal SSCs. The perichondral sheath is the original source of SSCs during initial bone formation. Remnants of stem cell-like populations can be found lining the periosteum of the bone collars throughout life where they have been shown to play important roles in regeneration and growth by preferentially generating chondrogenic and osteogenic, but not stromal or adipogenic, cell fates (Duchamp de Lageneste et al. 2018; Debnath et al. 2018; Ortinau et al. 2019; Murphy et al. 2020).

Although not definitively demonstrated yet, mSSCs are also present on the periosteum where they, upon injury, start expressing CD49f indicating an activated state with increased proliferative and osteogenic potential accommodating fracture repair (Marecic et al. 2015). Interestingly, cells enriched for SSC characteristics, including mSSC markers, localizing along the periosteum can also be labeled with Cathepsin K (CTSK), better known as an osteoclast marker, as early as E14.5 (Debnath et al. 2018). Due to the heterogeneous nature of CTSK-labeled cell populations with expression in multiple skeletogenic and non-skeletogenic cell types in bones, interpretations regarding cellular contributions and function drawn from results of a mouse model with CTSK driving a fluorescence reporter are difficult to interpret. Another study in mice using MX1(MX Dynamin Like GTPase 1)$^+$aSMA(Alpha Smooth Muscle Actin)$^+$ to mark a population enriching for periosteal SSCs along with the others has found a key role of these cells in bone defects (Ortinau et al. 2019). Overlap with CTSK was not investigated. A recent report even suggested additional SSC heterogeneity within the periosteum (Matthews et al. 2021). Alternatively, this might also underline the need for more specific SSC lineage tracing models. It remains to be determined if all SSC subsets originate from the initial populations of SSCs residing in the fibrous layer of the perichondrium during early development.

Perisinusoidal BMSCs have also been described as progenitors of osteogenic, chondrogenic, and adipogenic cell types albeit with varying extents depending on the markers and models used (Sacchetti et al. 2007; Morikawa et al. 2009; Park et al. 2012; Zhou et al. 2014; Ambrosi et al. 2017). Perivascular BMSCs in patient-derived skeletal tissue have been most efficiently labeled by STRO1, CD271 (NGFR), or CD146 (MCAM) (Simmons and Torok-Storb 1991, p. 1; Sacchetti et al. 2007; Tormin et al. 2011). In mice, CD140A/B (PDGFRA/B), SCA1, CD51 (ITGAV), CD105 (ENG), CD90 (THY1), and LepR are the best markers to enrich for SSCs associated with blood vessels (Koide et al. 2007; Morikawa et al. 2009; Park et al. 2012; Mabuchi et al. 2013a; Zhou et al. 2014). Specifically, a SSC-like population marked as SCA1$^+$PDGFRA$^+$CD24$^+$ characterizes a highly pure subendothelial cell population with bona fide stem cell characteristics including tri-lineage differentiation output in vitro and in vivo (Ambrosi et al. 2017). LepR-Cre mice crossed to tomato-fluorescence reporters have long been considered SSCs based on the fact that they label the majority of CFU-F forming cells of non-hematopoietic marrow (Ding et al. 2012; Zhou et al. 2014). It is important to remember that LepR-Cre has only been available with constitutive Cre-recombinase expression which leads to the labeling of the majority of osteogenic, chondrogenic, and adipogenic cells in the adult skeleton. A further caveat is that in addition to BMSCs, many other committed and mature cell types of the endothelial, mesenchymal, and even hematopoietic lineages express LepR. At the same time, SSC subsets of different skeletal regions express varying levels of LepR depending on their developmental state. This has now in part been mitigated by the generation of inducible LepR-CreER mice allowing a more time controlled investigation of LepR expressing cell contributions to skeletogenic tissue, albeit with remaining cellular heterogeneity (Sivaraj et al. 2021; Shu et al. 2021). The LepR-tracer mouse models therefore can give interesting insights into general skeletal biology but do not allow for specific conclusions of pure SSCs or even less SSC subpopulations. Finally, developmental origin is another contributor to SSC diversity. Cranial and vertebra stem cell populations localizing to the sutures of skull bones widely differ in their marker expression from SSCs in the appendicular skeleton but require more systematic interrogation and comparison in terms of lineage potential and molecular regulation (Zhao et al. 2015; Maruyama et al. 2016; Doro et al. 2017; Ransom et al. 2018; Shu et al. 2021). Altogether, these observations suggest that in contrast to the hierarchical model of HSC biology which relies on a single stem cell at the apex of the entire blood forming system, skeletogenic tissues encompass multiple SSC subsets with distinct trajectories of cellular output and function.

2.4 Downstream Lineages

SSCs are rare cell populations that give rise to more abundant committed downstream cell types of skeletogenic tissues. Each of these mature cell types fulfills specific functions important to the structural integrity of the skeleton. Classically, SSC biology has been treated as a black and white scenario of fixed bifurcation choices. However, recent discoveries have shifted that view dramatically. It is still believed that Paired-related homebox 1 (PRRX1) is the earliest as well as the most inclusive marker of skeletogenic tissues since virtually all mesenchymal cell types of long bones can be traced back to its expression (Logan et al. 2002; Ambrosi et al. 2017). The bone-forming osteoprogenitors, osteoblasts, and osteocytes are efficiently labeled by several markers such as RUNX2 (Runt-related transcription factor 2), GLI1 (Glioma-associated oncogene homolog 1), OSX (Osterix), or COL1A1 (Collagen, type I, alpha 1) while chondroprogenitors, chondroblasts, and chondrocytes are found within COL2A1 (Collagen, type 2, alpha 1) and Aggrecan expressing cells. Depending on the experimental approach and physiological setting, all SSC subtypes demonstrate osteochondrogenic potential, but not all SSC subtypes have been shown to give rise to stroma and/or fat. Seminal work that has identified a highly pure mouse SSC as the source to a transient bone cartilage stromal progenitor (BCSP) that is the origin of several osteogenic, chondrogenic, and stromal lineages with defined surface marker profiles never follows an adipogenic cell fate (Chan et al. 2015; Worthley et al. 2015). It provides, however, a stromal lineage positive for 6C3 expression that is strongly supportive of hematopoiesis in vitro and in vivo (Whitlock et al. 1987; Chan et al. 2013). Many SSC populations and their downstream stromal cell types also express high levels of HSC-supporting factors such as CXCL12 (Stromal cell-derived factor 1) and SCF (Kit-ligand) (Kfoury and Scadden 2015). Work on CXCL12-abundant reticular (CAR) cells, scattered throughout the marrow cavity in close proximity of sinusoid blood vessel, has supported these observations (Chan et al. 2009; Greenbaum et al. 2013). More recent work has found that specific periarteriolar osteolineage cells marked by LepR and Osteolectin expression provide a specialized niche for bone marrow lymphopoiesis (Shen et al. 2021). Therefore, both SSC and more lineage-restricted skeletal progenitor populations seem to be an integral part of bone marrow niches that control hematopoietic output.

The role of BMAT has been mostly ignored when studying skeletal biology in health and disease. Since aging bones lose skeletal mass and increase marrow fat, one of the prevailing determinants of bone health has been the ratio of osteogenic versus adipogenic differentiation from "MSCs". Given what is now known from SSCs at different anatomical locations and that BMAT accumulates in the medullary cavity, perisinusoidal SSCs specifically have been demonstrated to be the source of BMAT (Ambrosi et al. 2021b; Scheller et al. 2015; Ambrosi and Schulz 2017). Using a combination of lineage tracing models and transplantation experiments with prospectively FACS purified cell populations, the bone-resident adipogenic lineage could be tracked back to a perivascular SSC (Ambrosi et al. 2017). Committed adipogenic progenitors of the bone marrow can be detected by $SCA1^{+}CD24^{-}$ expression and further give rise to ZFP423-positive pre-adipocytes. Aging and obesity leads to increased prevalence of adipogenic cell types which impair regenerative processes of the skeletal and hematopoietic systems. Other independent studies employing different markers, i.e., RANKL, PREF1, and Adiponectin, also found a negative correlation between the number of adipogenic cells present and maintenance of balanced bone remodeling and hematopoiesis (Fan et al. 2017; Zhong et al. 2020). Interestingly, several single RNA-sequencing approaches have identified that a large proportion of bone-resident mesenchymal progenitor cells exhibit an adipogenic phenotype based on their expression of markers commonly related to adipose tissue. These pericyte-like cells express many vasculature, hematopoiesis, and skeleton modulating signals and may play a much more important role in systemic regulation than previously thought (Cawthorn et al. 2014; Ambrosi et al. 2017; Tikhonova et al. 2019; Zhong et al. 2020; Yu et al. 2020). Future studies are needed to examine the extent to which different SSC subtypes and their downstream cell populations cross-communicate to control cellular demand for maintaining an intact skeletal structure.

2.5 *Current Shortcomings of SSC Biology*

It has to be emphasized that bone-forming SSCs normally do not circulate through the blood stream and that they have limited migratory potential to begin with (Chan et al. 2015; Murphy et al. 2020). Unlike HSCs, the limited homing potential of SSCs may be a reason for the missing success of translational efforts using "MSCs" for regenerative therapies. Additionally, the heterogeneous nature of "MSC" populations only provides a small number of therapeutical active SSCs insufficient for treating bone disorders. Furthermore, using more accessible sources of "MSCs", e.g., adipose tissue, muscle, cord blood, that only display skeletogenic capacity under highly artificial or conducive conditions are unlikely to succeed in an in vivo setting that lacks these prerequisites (Reinisch et al. 2015; Sacchetti et al. 2016; Galipeau and Sensébé

2018). In a more general view, systemic delivery of stem cell enriched populations for a wide array of diseases has been tried and the majority have failed in experimental and clinical trials. While "MSCs" are immunologically inert, the majority of intravenously administered cells becomes trapped in the lungs of recipients (Fischer et al. 2009). The fraction of migrating cells, likely not bona fide stem cells, arriving in the target organ does not accomplish the desired regeneration but rather mediates partial disease alleviation through paracrine modulation, such as anti-inflammatory actions (Bianco et al. 2013). In summary, the knowledge we have gained about bone-resident bona fide stem cells should be applied to thoroughly think of more effective ways to use SSCs for feasible translational approaches. Cell populations need to be investigated more stringently using in vivo assays (Bianco et al. 2013; Ambrosi et al. 2019). Identifying pure SSC populations will help to learn more about factors and signals involved in normal and diseased SSC niches. We will also have to assess whether fine-tuning SSCs in their endogenous microenvironment through the delivery of niche-modulating factors might be more effective than extracting cell populations for ex vivo expansion and manipulations before local transplantation. Modeling of the bone marrow environment in particular using human-derived cells will help us understand basic cellular and molecular components of defined bone marrow microenvironments paving the way to new vantage points for developing novel therapeutic strategies for a broad number of skeletal and hematopoietic diseases.

3 Skeletal Stem Cells and the Niche

3.1 Niche Formation

Around the same time first experimental proof for the existence of a HSC had been gathered and leading up to the report of a colony forming stromal cell, experiments by Tavassoli and Crosby could show that boneless marrow pieces have the ability to recreate a hematopoiesis-supporting organoid complete with bone, cartilage, and stroma, when transplanted at extramedullary sites (Tavassoli and Crosby 1968). In line with that, Schofield later postulated the existence of a stem cell niche where a stem cell interacts with other cells that determine its behavior to either remain quiescent, self-renew, or replenish tissue with downstream cell types as demand requires it (Schofield 1978). It took almost another 40 years until it was unequivocally demonstrated that the hematopoietic stem cell niche indeed was a combination of two stem cell systems; HSCs and multipotent skeletal stem and progenitor cells that are crucial for initiation and maintenance of such a microenvironment (Chan et al. 2013, 2015). Assays that deviated from the commonly purified SSC populations or in vitro expanded colonies from plastic-adherent cultures enriched for stem cells had to be developed. Subcutaneous as well as renal capsule sites as preferred anatomical regions for transplants of freshly purified cell populations have emerged as they are highly vascularized allowing proper engraftment of cells. Upon transplantation, SSCs initially proliferate and then commit to cartilage fates to undergo endochondral ossification to form bone upon blood vessel invasion and to create a stromal layer that attracts circulating host-derived HSCs (Chan et al. 2009). If the same experiments are performed with more committed skeletal cell types with a lack of stem cell characteristics, no hematopoietic niche is formed but rather bony ossicles or pure cartilage occurs (Chan et al. 2013, 2015; Debnath et al. 2018). This infers that multiple skeletogenic lineages derived from SSCs are necessary to establish hematopoiesis as well as the need to replenish those cell types through a multipotent stem cell population. SSCs might further dictate lineage output by providing autocrine and paracrine signaling as a way to organize tissue structure. For example, it has now been shown that SSCs of the growth plate maintain their stem cell status by expression of Wnt-inhibitory signals and that activation by microenvironmental factors for Wnt and other signaling pathways dictates differentiation (Ambrosi et al. 2021b; Hallett et al. 2021). The emergence of extramedullary hematopoiesis in SSC-derived ossicles might therefore be a miniature model of the processes underlying niche formation in the native bone marrow.

SSCs and HSCs as well as their downstream cell types together with endothelial and nerve cells maintain intra- and inter-cell type communication thereby orchestrating niche function. As elaborated on, this occurs on the level of reciprocal signaling axes with specific ligand-receptor interactions. Endochondral ossification by SSCs is a prerequisite for bone marrow niche formation (Chan et al. 2009). HSC niches are established after embryonic day 15.5 when fetal liver HSCs start circulating to home to the bone marrow (Rowe et al. 2016). Influx of HSCs is enabled by the forming blood vessel network first infiltrating into the diaphyseal primary ossification center of long bones. Osterix expression in SSCs and downstream cell populations is crucial for niche initiation and HSC engraftment (Chan et al. 2009; Mizoguchi et al. 2014; Coşkun et al. 2014). The early hematopoietic niche in the fetal liver and embryonic bone marrow promotes high proliferative HSC activity, while the postnatal marrow niche primarily upholds quiescence (Morrison et al. 1995; Christensen et al. 2004). HSCs remain rare during adulthood and difficult to identify in tissue sections. Systematic in situ labeling finds hematopoietic stem and progenitor cells (HSPCs) throughout the medullary cavity most frequently in close proximity to sinusoidal blood vessels adjacent to LepR and CXCL12 expressing stroma and infrequent close to arteries, Schwann cells, adipocytes, and megakaryocytes (Kokkaliaris et al. 2020). Spatially distinct niches with variation in signals promoting quiescence versus activation of adult HSCs have been suggested. Specifically, endosteal and arteriolar spaces have been thought to be important anatomical landmarks for niches with cycling HSCs, however, whole tissue clearing and deep confocal imaging techniques have shown that the majority of dividing and non-dividing HSCs are found in perisinusoidal localizations (Kiel et al. 2005; Ding et al. 2012; Kunisaki et al. 2013; Acar et al. 2015). Discrepancies in part stem from the markers used to identify HSPCs and overall experimental settings such as irradiation conditioned reconstitution models. The investigated bone sites might similarly influence readouts, as a recent study using live animal imaging of HSPCs in calvarial bone marrow found long-term HSCs residing close to both sinusoidal blood vessels and the endosteal surface (Christodoulou et al. 2020). Despite the technical deviations, it is reasonable to assume that multiple niches with distinct cellular compositions and functions exist.

3.2 From Skeletal Stem Cell to Niche

SSCs as the starting point for bone and niche assembly have made cells of the osteolineage major candidates for providing HSC niche signals including SCF and CXCL12 (Méndez-Ferrer et al. 2010; Greenbaum et al. 2013). This was supported by findings that increases in osteoprogenitors and bone mass correlated with an elevated presence of HSPCs (Taichman and Emerson 1994; Calvi et al. 2003; Zhang et al. 2003). However, these results might have been misinterpreted by employing markers that overlap in osteoblastic and stromal cell types. It is evident that high skeletal density only promotes HSPC function until it reaches a threshold that limits bone marrow area leading to ectopic hematopoiesis. For instance, mouse models of osteopetrosis show increased loss of medullary space corresponding to enhanced extramedullary hematopoiesis including enlarged spleens with active hematopoiesis (Pasold et al. 2013; Herber et al. 2019). Stromal and vascular cell types, rather than osteoprogenitors, have also been considered crucial for adult HSC niches since they express the highest levels of HSPC maintenance signals, and specific loss of the same signals from these cells depletes HSCs (Ding et al. 2012; Ding and Morrison 2013). Interestingly, CXCL12 derived from the osteolineage seems to be important to maintain early lymphoid progenitors, underlining that differential cellular contributions and signaling axes characterize varies niches (Katayama et al. 2006; Ding and Morrison 2013). Many markers have been used to label stromal stem and progenitor cells that express crucial factors for HSPC maintenance. CXCL12-eGFP reporter mice ubiquitously mark an abundance of reticular cells (CAR-cells) that line blood vessels and are found adjacent to HSPCs (Sugiyama and Nagasawa 2012; Greenbaum et al. 2013). While LepR labeling remains highly heterogeneous for reticular, osteoprogenitor, and mature skeletal cells, it does appear to enrich for types of HSPC-supportive niche cells (Zhou et al. 2014). For instance, LepR but not Nestin expressing cells play a determining role in their contributions to maintain niche integrity (Méndez-Ferrer et al. 2010; Ding et al. 2012). LepR and Parathyroid receptor expressing lineages also adapt and reshape niches in response to systemic metabolic demands (Yue et al. 2016; Fan et al. 2017). One study found that in a regenerative setting Adiponectin expressing bone marrow adipocytes promote hematopoietic recovery after ablation by secreting SCF (Zhou et al. 2017). Whereas adipocytes are abundant after myeloablation, a small populations of Adiponectin positive stromal cells might rather be the main source of SCF mediating the observed effects (Zhong et al. 2020). Conversely, there is also evidence that BMAT has a negative effect on HSC engraftment and expansion which has provided evidence for a direct function of BMAT in hematopoiesis under homeostatic conditions (Naveiras et al. 2009). In summary, new genetic models and experimental systems that offers greater cellular specificity are needed to more accurately dissect how distinct SSC types and smaller subsets of SSC downstream populations contribute to regulation of hematopoiesis.

Finally, the hematopoietic supportive effect of SSC-derived lineages must also be considered in the context of other cell types and cell lineages. The perivascular anatomical localization of some niche cells implies a role for endothelial cell types (Ding and Morrison 2013; Itkin et al. 2016). SSCs promote vascular infiltration through expression of VEGF (Chan et al. 2015). Conversely, it has been reported that endothelial cells express high levels of Notch-ligands to stimulate their own proliferation but also to direct SSC lineage cell output and behavior thereby controlling HSPC biology (Ramasamy et al. 2016; Lampreia et al. 2017). Loss of the vascular network impairs bone marrow microenvironments and thus HSC function (Kusumbe et al. 2016). Aside from BMSCs, macrophages reside along the endothelium and express VCAM1 (Vascular cell adhesion molecule 1) guiding homing and retention of HSPCs (Mabuchi et al. 2013b; Li et al. 2018). Another hematopoietic cell type playing a part in stem cell niches are megakaryocytes. These cells have been shown to express factors directly and indirectly influencing niche dynamics, in particular paracrine factors like TPO (Thrombopoietin), TGFB1 (Transforming Growth Factor Beta 1), and CXCL4 (Platelet Factor 4) (Bruns et al. 2014; Zhao et al. 2014; Nakamura-Ishizu et al. 2014). Moreover, long bones become innervated around mouse embryonic day 15, about the same time blood vessels infiltrate and endochondral ossification creates the microenvironment allowing HSCs to relocate to marrow sites (Bidegain et al. 1995; Sisask et al. 2013). Data shows that sympathetic nerves in the form of non-myelinating Schwann (glial) cells can regulate HSPCs by co-localization and expression of niche factors, again indirectly through regulation of osteogenic and stromal lineage cells (Katayama et al. 2006; Yamazaki et al. 2011). In contrast, pain receptor neurons extending into the bone marrow allow granulocyte colony-stimulating factor (GCSF)-induced HSPC egress via direct actions by secretion of calcitonin gene-related peptide (CGRP) as an adaptive mechanism to start an immune response to avoid organ damage (Gao et al. 2020). In summary, the accumulative work of many research groups underscores the complexity of SSC-derived HSC niches with its diverse cell types of different developmental origins and maturation stages (Figs. 1 and 2). Distinct niches exist to serve specific purposes that are controlled through direct cell–cell contact and ligand-receptor interactions, however, relevance of many of these observations to the human setting remain to be explored.

3.3 Niches During Aging and Disease

Given the many cells involved in supporting proper niche function and evidence that ablation of single factors from specific subpopulations can impair skeletogenic and hematopoietic processes, it is not surprising that microenvironmental alterations in the niche can drive the development of hematopoietic diseases. Multiple studies have shown that niche changes could be genotoxic and oncogenic. On the other hand, there is also credence to a mechanism whereby malignant stem and progenitor cells transform healthy niches for their advantage to expand. Both processes would suggest that autocrine and paracrine factors establish and determine altered actions by the various cell types surrounding the stem cell systems. One well-studied example is the aging process of hematopoiesis. Many drivers of HSC aging have been shown to be intrinsic to stem cells (Beerman et al. 2010). This includes mutagenesis and epigenetic alterations promoting increased self-renewal and skewed myeloid lineage output (Rossi et al. 2005; Pang et al. 2011; Beerman and Rossi 2015). Somatic mutations in the hematopoietic compartment can affect fitness conferring selective growth advantage in clonal populations. Those mutations are more frequent in aged populations where they perpetuate malignancy which is now known under the clinical term of Clonal hematopoiesis of indeterminate potential (CHIP) (Jaiswal and Ebert 2019). Inversely, the development of features such as increased expansion, myeloid skewing, and CHIP in HSPCs could also be caused by selective pressure from an altered microenvironmental signaling. This suggests that extrinsic factors of different niche cell types might facilitate the initiation and progression of malignant HSPC clones.

The reciprocal co-dependence of bone health to healthy hematopoiesis is clearly evident in observations that aberrant hematopoiesis drives loss of bone mass and diseased skeletogenesis debilitates hematopoietic activity (Raaijmakers et al. 2010; McClune and Majhail 2013). Adipogenic cells also accumulate in skeletal tissues during aging, obesity, and disease, and are inversely correlated with bone mass, i.e., the prevalence of osteochondrogenic cell types (Ambrosi et al. 2017). Bone marrow adipocytes are also negative regulators of hematopoiesis (Naveiras et al. 2009). The secreted drug targetable molecule Dipeptidyl peptidase IV (DPP4) expressed by adipocytes impairs hematopoiesis by cleaving CXCL12 as well as a number of pro-osteogenic factors (Broxmeyer et al. 2012; Ambrosi et al. 2017). A recent study additionally suggests that bone marrow resident adipogenic lineage cells at least in part contribute to myeloid skewing, a hallmark of HSPC aging, specifically by secreting pro-osteoclastic factors such as RANKL (Receptor Activator of NF-κB Ligand) that promote bone resorption (Yu et al. 2020). SSC-derived stroma has also been implied to elevate osteoclast activity through CSF1 (Colony Stimulating Factor 1) and RANKL expression (Cao et al. 2005; Nakashima et al. 2011). Paradoxically, the diminished skeletogenic activity of aged SSCs, which predominantly give rise to myeloid-supportive stromal subpopulations, simultaneously promotes enhanced osteoclastogenesis through high expression of pro-inflammatory cytokines, including CSF1 (Ambrosi et al. 2021a). Consequently, age-related skeletal loss and hematopoietic derangement are exacerbated by combined mechanisms of reduced bone formation, increased adipogenesis, and higher bone resorption originating from both distorted SSC and HSC activity, likely including age-related changes to the niche (Fig. 2). Earlier studies indicated that exposing aged bone marrow to a young blood circulation or aged HSCs to a young microenvironment by transplantation into young mice could, at least partially, rejuvenate niche functions (Baht et al. 2015; Kuribayashi et al. 2019). More recent work, however, challenges these observations claiming that exposure to a young blood circulation fails to rejuvenate aged

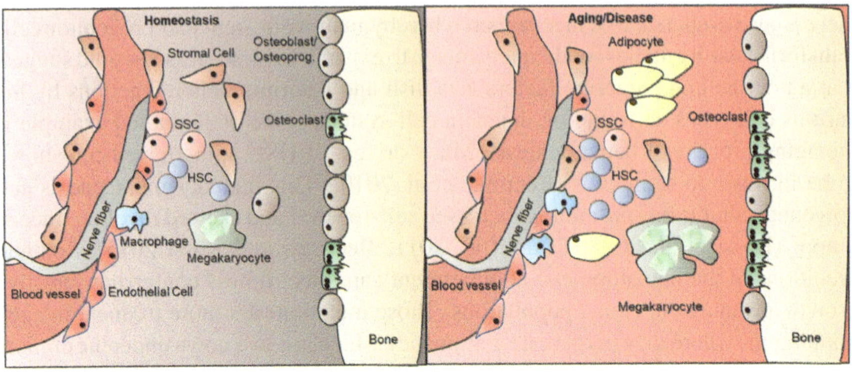

Fig. 2 The hematopoietic stem cell niche in homeostasis and disease/aging. Left, homeostatic HSC niche showing HSCs in proximity to blood vessels surrounded by key cell types that secrete regulatory factors. Right, aging and disease alters cellular composition of HSC niches leading to higher HSC prevalence and altered lineage output. HSC aging can lead to niche changes or cellular changes in the microenvironment can lead to HSC aging

HSCs and SSCs (Ho et al. 2021; Ambrosi et al. 2021a). This suggests that there might be a concentration dependent response of niche cells to microenvironmental signals that in an aged setting has to exceed physiological levels. Moreover, endosteal niches, which are thought to support lymphopoiesis, are lost during age-related bone marrow remodeling (Ho et al. 2019). As a result, a β2-adrenergic to inflammatory IL6 axis promotes megakaryocyte differentiation elevating myeloid lineage count and boosting HSPC expansion. Aged hematopoietic cells also develop a senescence-associated secretory phenotype (SASP) that promotes "inflammaging". Removal of senescent cells with senolytic agents therefore represents a possible strategy to rescue bone loss and potentially hematopoiesis during aging (Farr et al. 2017). It remains to be determined whether elimination of specific cell types such as cells of the myeloid or mesenchymal lineages are responsible for the beneficial effects since current senolytics are not cell-type specific.

Local and systemic persistence of pro-inflammatory cytokine signaling can also drive SSC dysfunction either directly or indirectly through HSPCs (Mitroulis et al. 2018; Josephson et al. 2019). High levels of GCSF primarily from vascular niches boost HSC cell cycle entry and mobilization (Boettcher et al. 2014). GCSF inhibits osteogenic activity and reduces the expression of stem cell retention molecules such as CXCL12 (Semerad et al. 2005). Inflammatory states also promote the conversion or development of abnormal niches that may facilitate initiation and progression of leukemia. While interleukins and interferons are key mediators of inflammation, they are also increased in myelodysplastic syndrome (MDS) (Fozza et al. 2016). Chronic IL1, for instance, accelerates cell division and myeloid differentiation of HSPCs by early activation of a gene program depending on the transcription factor PU.1 (Pietras et al. 2016). Similarly, persistent type 1 and 2 interferon signaling keeps HSPCs in an activated non-dormant state that favors myeloid-bias (Essers et al. 2009;

Matatall et al. 2014). MDS derived BMSCs have been shown to display disturbed differentiation with upregulation of genes involved in cell adhesion and angiogenesis (Medyouf et al. 2014). Thus, SSC-derived bone marrow stroma seems also likely to be one of the main sources of inflammatory signals promoting MDS. Indeed, active NF-κB signaling in BMSCs fosters strong transcriptional overexpression of negative regulators of hematopoiesis, many of which are involved in inflammatory signaling (Ping et al. 2019). The fact that myelodysplasia is a clonal hematopoietic disease and that it can develop as a consequence of the loss of a single gene in osteoprogenitors or stroma demonstrates that both intrinsic as well as microenvironmental factors in the niche can initiate disease (Raaijmakers et al. 2010). Finally, the bone marrow is a preferred site for metastasis of cancers, specifically breast and prostate cancer. Circulating disseminated tumor cells have been suggested to compete with HSCs for local perivascular niches in order to promote their own survival and growth (Sipkins et al. 2005; Shiozawa et al. 2011). Malignant cells might respond to the same trophic and atrophic signals that normally regulate HSC thereby allowing them to "hijack" the niche in agreement to the "seed and soil" theory of metastatic spread (Fidler 2003). Understanding these niche signals might therefore reveal new opportunities for targeting bone metastasis.

3.4 Molecular Targeting of the Niche

Recent studies modeling SSC-specific lineage output by altering signaling cues in the environment have found key molecular pathways for maintaining skeletal integrity, among them Hedgehog, BMP (Bone Morphogenic Protein), and VEGF (Vascular endothelial growth factor) signaling (Chan et al. 2015; Mizuhashi et al. 2018; Newton et al. 2019). For example, diabetic mice and humans show a diminished regenerative potential of the skeleton which has been found to be based on the loss of Indian Hedgehog (IHH) signaling in the skeletal stem and progenitor cell compartment restorable by exogeneous IHH (Tevlin et al. 2017). Similarly, aging of SSCs promotes anti-osteogenic, pro-apoptotic, pro-inflammatory features that can be reversed by re-introducing lost Sirtuin1 expression, pharmacologic inhibition of NF-κB activation, or combined SSC-directed BMP2 stimulation with blockade of CSF1 signaling on monocytic cell lineages (Josephson et al. 2019; Ambrosi et al. 2020, 2021a). If left untreated, intrinsic and extrinsic factors during aging and disease change cell signaling and skew lineage output of SSCs. Osteoarthritis of the joints stems from degeneration of the chondrogenic articular sites that are replaced with osteogenic tissue, originating from the same stem cell source. Under normal circumstances, cartilage possesses poor regenerative ability. Introducing microfractures into the areas where articular cartilage usually resides activates local SSCs and additional space-restricted delivery of BMP2 protein and anti-VEGF antibody can then guide robust formation of endogenous hyaline cartilage (Murphy et al. 2020). In the future, identifying factors that naturally activate dormant SSCs could be combined with cartilage inducing compounds using injectable scaffolds to achieve minimal invasiveness during treatment (Li et al. 2019). As described above, alterations in paracrine profiles of SSC lineage cells are an additional determinant of niche cell behavior. For example, hematopoietic cell types are controlled by TGFb, a factor mainly produced by SSC-derived mesenchymal as well as endothelial niche cell types (Blank and Karlsson 2015). Studies in mice have shown that maintaining young TGFb1 levels could rebalance myeloid-skewed lineage output by HSCs during aging (Challen et al. 2010; Quéré et al. 2014). This could be challenging though as TGFb needs to be under strict dose and spatiotemporal control as it also could drive adverse fibrosis (Agarwal et al. 2016). Finally, counteracting elevated tumor necrosis factor (TNF) expression in pro-inflammatory stromal cell subsets seems to be a feasible strategy to rescue inflammation-mediated niche impairments (Mead et al. 2017). Thus, preventing the development of an inflammatory bone marrow environment is crucial to skeletal integrity and deterence of malignant transformation (Leimkuehler et al. 2021). Targeting SSC lineage cells in their endogenous environment to reactivate intrinsic properties and regulate paracrine actions could be a more promising and feasible approach than using transplants of in vitro expanded autologous or even less suitable allogenic stem and progenitor populations.

Novel methods to treat niche defects are beginning to appear that build on increasingly detailed dissection of healthy and neoplastic bone marrow microenvironments using single cell approaches. However, a major limitation of functional studies of the hematopoietic niche remains that the major in vivo functional assays for stem cell engraftment rely on cytotoxic radio- or chemotherapy-based methods for clearing the niche. This type of pre-conditioning severely disrupts normal signals and the cell composition of the niche. Alternate methods employing antibody conditioning now present the possibility of highly specific ablation of HSPCs with much reduced collateral damage to the niche, thus revealing more of the normal signals and cell types directing stem cell niche functions (Czechowicz et al. 2007; Palchaudhuri et al. 2016). Antibody conditioning to deplete specific cellular subsets could now be combined with novel transgenic mice utilizing Crispr-mediated gene activation or silencing for highly specific and sophisticated interrogation of genes and distinct cell types in network analysis of the niche. At the same time, antibody-mediated conditioning presents an exciting new strategy for highly specific clinical myeloablation that circumvents many of the most serious side effects in broad toxic treatment regimens (Agarwal et al. 2019). In the setting of allogenic cell transplantation, niche dynamics are not only altered by pre-conditioning but also potential graft versus host diseases (GvHD) due to HLA mismatches. This has been shown to lead to increased bone loss with higher incidence of fractures and osteoporosis (McClune and Majhail 2013; Pundole et al. 2015; Schwarz et al. 2021). New approaches to specifically alter the niche using antibody-conjugated drug therapy thus offers exciting possibilities for both in situ niche research as well as translational applications for gene therapy of the hematopoietic system (George et al. 2019).

3.5 Niche Engineering

Recreating stem cell niches from their basic cellular and molecular constituents can help identify and model the primary mechanisms that enable different stem cell niches in complex tissues to coordinate their activity with each other. The bone marrow niche in particular offers a glimpse into how two distinct stem cell systems, SSCs and HSCs interact to facilitate skeletogenesis and hematopoiesis and how derangements in their signaling crosstalk lead to disease. Early genetic studies in mice have identified specific signaling pathways and genetic regulators that are necessary for HSC niche formation. For instance, Osterix expression from skeletal stem and progenitor cells is necessary to initiate niche formation, whereas SCF is not needed to establish hematopoiesis (Chan et al. 2009). Similarly, VEGF signaling by skeletal stem and progenitors is also critical as antagonizing vascularization through VEGF blockade suppresses niche formation entirely.

Although potent combinations of signaling factors that could activate SSC and HSC function to generate niche tissue de novo have been already discovered and successfully applied to regenerative settings (Chan et al. 2009; Tevlin et al. 2017; Murphy et al. 2020), new less invasive approaches to deliver these signals with greater precision for enhanced regenerative responses have to be devised to spare patients from surgical intervention. The complexity, density, and at many locations avascularity of bone tissue represents a particular challenge. In situ manipulation of site-specific signaling cascades are emerging from the field of immune engineering using nanobodies and Chimeric antigen receptor (CAR) immune cell therapy. Camelid single-domain antibody fragments, or nanobodies, are a novel promising class of therapeutics. Unlike the larger antibodies, nanobodies easily penetrate the cytosol of living cells, diffuse even through very dense tissue, and are rapidly cleared from the blood stream thereby reducing off-target effects (Jovčevska and Muyldermans 2020). These unique features plus their FDA approval make nanobodies an ideal tool for the use in skeletal tissues, e.g., as a blocker of aberrant niche signals. A yeast library-based in vitro platform can facilitate the production of high affinity nanobodies for virtually any given niche factor (McMahon et al. 2018). Injections of one or a combination of nanobodies could then be used to activate and/or block signaling to allow timed and space-restricted niche manipulation.

Other studies focused on characterizing the role of specific extracellular matrices. To optimize the delivery of growth factors, antibodies, or small molecules and translate findings to pre-clinical animal models and human patients, fine-tuning and scaling up of existing systems are needed. Importantly, since hematopoiesis, skeletal formation, and regeneration are expedited through a series of sequential events that are coordinated in a spatiotemporal manner, the new materials and systems need to overcome caveats such as lack of specificity to target site and uncontrolled release kinetics. Biomaterial sciences have generated promising types of scaffolds that can be readily placed to local sites in a regenerative setting. If intact niche structure with its cells and extracellular matrix (ECM) components that retains growth factors, biomechanics, and structural properties are lost, endogenous or synthetic materials are needed for repair. Researchers already have used decellularized matrix to provide tunable 3D-scaffolds modeling human SSC-HSC niches in vitro and to promote bone healing in vivo (Lai et al. 2013; Emami et al. 2020). Prevailing disadvantages are that in vitro generated decellularized ECMs tend to poorly represent native organization, and that existing protocols to prepare tissue derived decellularized ECMs fail to maintain mechanical properties due to harsh treatment, and if using dense tissues like bone are hard to re-cellularize (Kim et al. 2018). Synthetic materials can deliver growth factors loaded to scaffolds by entrapment, adsorption, biotinylation, or covalent binding. The most widely used scaffolds have been made of monolithic biomaterials which release a single compatible factor by diffusion. For instance, polymers, e.g., Poly(lactic-co-glycolic acid) (PLGA), can be employed for local retention with partially timed control of bioactivity, with the strongest release initially. Subcutaneous PLGA scaffolds loaded with CXCL12 and transplanted subcutaneously into mice promote local mesenchymal and hematopoietic cell engraftment (Thevenot et al. 2010). Polysaccharides are attractive as they are natural polymers that are highly abundant in nature, non-toxic, stable, low cost, and biocompatible. They possess hydrophilic groups endowing them with charges in solution and dividing them into cationic (e.g., chitosan) and anionic (e.g., alginate) polymers. Growth factors can be transiently linked by charges mimicking high affinity of growth factors to endogenous ECM. This allows for the incorporation of multiple factors into one scaffold with a better spatiotemporal release (Rajam et al. 2011). Nonetheless, it needs to be considered that ionic tethering includes non-specific charging with local proteins when introduced to an in vivo setting which might interfere with loaded factors and that bonding of scaffolds to host bone tissue requires additional osteoconductive particles such as hydroxyapatite and diopside (Saravanan et al. 2016). Altogether, to find the most suitable biomaterials to engineer skeletal stem cell niches and reverse disease-induced tissue changes, scaffolds will have to be individually optimized for their purpose. This will likely entail the use of a combination of materials, chemicals, and 3D-bioprinting approaches.

Acellular scaffolds, such as decellularized matrices or synthetic polymers, have been observed to promote ossicle formation but are insufficient without complimentary cell seeding (Bourgine et al. 2014; Ventura Ferreira et al. 2016). In particular, different types of matrices have been observed to potentiate humanized ossicles formation in xenograft settings. Such ossicles significantly improve engraftment of both normal and malignant hematopoietic progenitors for analysis in immunodeficient mice as mouse bone marrow niches cannot fully host normal human hematopoiesis by transplanted human HSPCs. In vitro expanded human BMSCs can be packed into an extracellular matrix and subcutaneously transplanted into immunodeficient mice. Initial PTH treatment of recipient mice is necessary to induce hematopoiesis-supporting ossicle formation through endochondral ossification (Reinisch et al. 2017). Once ossicles have successfully formed, the hematopoietic system of the mouse could be ablated and reconstituted with patient-derived HSPCs either intravenously or by directly injecting them into the marrow cavities of ossicles. Alternatively, BMSCs can be seeded onto three-dimensional scaffolds and differentiated to form hypertrophic chondrocytes in vitro before transplantation under the skin of mice which will also allow sufficient host-derived ingress of endothelial and blood cells to establish a functional niche (Bourgine et al. 2014). Co-engrafting both human skeletal and hematopoetic lineages allow for a more functional recapitulation of human-like microenvironments to investigate leukemias and other hematological malignancies such as myelofibrosis. Further, heterotopic solid tumor models in mice harboring functional ossicles can be established to study metastasis into stem cell niches. More sophisticated approaches now strive to engineer human bone marrow proxies ex vivo on microfluidic chips that could recreate an entire stem cell niche with purely human cells (Sieber et al. 2018). This method not only provides a potential alternative to animal research but also allows investigations of minimal perturbations in a highly controlled setting. However, broader utility of these "niches-on-a-chip" is currently limited and restricted by batch-to-batch variations of BMSCs underlining the need for better characterization of niche-initiating SSC populations (Bourgine et al. 2018). Understanding signaling crosstalk in human niches would be incomplete without a detailed analysis of the human SSC. Fortunately, great progress has been made in characterizing specific human SSC populations (Sacchetti et al. 2007; Chan et al. 2018). Reinstating proper SSC function, i.e., activation to proliferate and differentiate as desired, might be sufficient to not only correct skeletal disorders but also to reprogram aberrant HSC function and lineage output. Studies focusing on tuning SSC dynamics will yield new therapeutic opportunities for age- and disease-related skeletal disorders.

Recent advances in cancer cell therapy technologies support the feasibility of a strategy to engineer cells of hematopoietic lineages to express and release appropriate regenerative factors to specific sites in response to externally applied focused energy sources (Feins et al. 2019; Gamboa et al. 2020; Klichinsky et al. 2020). Using gene editing techniques, expression cassettes placing synthesis of crucial niche factors such as BMP2 or CXCL12 under control of heat-shock or magneto-responsive elements can be introduced into patient-derived hematopoietic cell types such as monocytes and T-cells (Andersson et al. 2014). Engineered stimulus-responsive cells can be further endowed with specific homing and engraftment capabilities by engineering them to express CARs that recognize specific cell antigens expressed by distinct skeletal niche cell types. This will make these engineered hematopoietic cells ideal delivery vehicles for regenerative factors, as they could be introduced locally as well as intravenously to home to affected tissues. Administration of spatiotemporally directed pulses of focused heat or magnetic fields to specific areas can induce highly localized production and secretion of regenerative factors to skeletal niches for tissue regeneration. In sum, combining basic niche biology research that provides insight into key cell populations and molecular targets with cutting-edge ways to spatiotemporally deliver therapeutic factors will shape the future treatment paradigms of SSC-HSC niche related diseases.

4 Conclusions

Bone marrow stem cell niches are integral to the maintenance and regeneration of skeletogenesis and hematopoiesis. They are composed of cells with multiple developmental origins ranging from mesenchymal, endothelial, hematopoietic to neuronal lineages and found in distinct localizations from endosteal to perivascular regions (Pinho and Frenette 2019). Autocrine and paracrine signaling axes between progenitor and mature cell types determine cellular behavior and output. SSCs create and shape the space for hematopoiesis in the medullary cavity (Chan et al. 2009, 2015). Together with their downstream cell populations, they play a major part in balancing hematopoietic demand. The prevalent classic perception of a single SSC type as the source of bone, cartilage, fat, and stroma appears to be outdated. Emerging data rather suggest the existence of diverse SSC subtypes with distinct anatomical neighborhoods and restricted differentiation capacities that fulfill defined purposes (Debnath et al. 2018; Mizuhashi et al. 2018; Murphy et al. 2020). For example, in lieu of the osteogenic versus adipogenic bifurcation choice of a ubiquitous SSC, the abundance and stimulation of an osteochondrogenic SSC subtype that facilitates bone formation compared to a perivascular hematopoiesis supporting adipogenic SSC subtype might determine skeletal health (Chan et al. 2015; Ambrosi et al. 2017, 2021b).

While the SSC compartment acts as the "soil", HSCs represent the "seeds" that are nourished. Small alterations in niche cells have the ability to shift dynamics and promote disease. Similarly, HSC-intrinsic changes can drive clonal hematopoietic malignancies transforming the microenvironment along the way (Beerman et al. 2010; Jaiswal et al. 2014). The fields of SSC and HSC biology have long been studied as single entities. Emphasizing the research of niches as a whole will guide a better understanding of cellular architecture and communication. Translating novel findings of cellular structures and spatiotemporal signaling by coupling them to engineering approaches from the fields of biomaterials and immune cell therapy will allow the development of paradigm shifting therapies for skeletal and hematological diseases and beyond.

References

Acar M, Kocherlakota KS, Murphy MM et al (2015) Deep imaging of bone marrow shows non-dividing stem cells are mainly perisinusoidal. Nature 526:126–130. https://doi.org/10.1038/nature15250

Agarwal A, Morrone K, Bartenstein M et al (2016) Bone marrow fibrosis in primary myelofibrosis: pathogenic mechanisms and the role of TGF-β. Stem Cell Investig 3. https://doi.org/10.3978/j.issn.2306-9759.2016.02.03

Agarwal R, Dvorak CC, Prohaska S et al (2019) Toxicity-free hematopoietic stem cell engraftment achieved with anti-CD117 monoclonal antibody conditioning. Biol Blood Marrow Transplant 25:S92. https://doi.org/10.1016/j.bbmt.2018.12.172

Ambrosi TH, Goodnough LH, Steininger HM et al (2020) Geriatric fragility fractures are associated with a human skeletal stem cell defect. Aging Cell 19:e13164. https://doi.org/10.1111/acel.13164

Ambrosi TH, Longaker MT, Chan CKF (2019) A revised perspective of skeletal stem cell biology. Front Cell Dev Biol 7:189. https://doi.org/10.3389/fcell.2019.00189
Ambrosi TH, Marecic O, McArdle A et al (2021a) Aged skeletal stem cells generate an inflammatory degenerative niche. Nature 1–7. https://doi.org/10.1038/s41586-021-03795-7
Ambrosi TH, Sinha R, Steininger HM et al (2021b) Distinct skeletal stem cell types orchestrate long bone skeletogenesis. elife 10:e66063. https://doi.org/10.7554/eLife.66063
Ambrosi TH, Schulz TJ (2017) The emerging role of bone marrow adipose tissue in bone health and dysfunction. J Mol Med 95:1291–1301. https://doi.org/10.1007/s00109-017-1604-7
Ambrosi TH, Scialdone A, Graja A et al (2017) Adipocyte accumulation in the bone marrow during obesity and aging impairs stem cell-based hematopoietic and bone regeneration. Cell Stem Cell 20:771-784.e6. https://doi.org/10.1016/j.stem.2017.02.009
Andersson HA, Kim Y-S, O'Neill BE et al (2014) HSP70 promoter-driven activation of gene expression for immunotherapy using gold nanorods and near infrared light. Vaccines (basel) 2:216–227. https://doi.org/10.3390/vaccines2020216
Baht GS, Silkstone D, Vi L et al (2015) Exposure to a youthful circulation rejuvenates bone repair through modulation of β-catenin. Nat Commun 6:7131. https://doi.org/10.1038/ncomms8131
Baryawno N, Przybylski D, Kowalczyk MS et al (2019) A cellular taxonomy of the bone marrow stroma in homeostasis and leukemia. Cell 177:1915–1932.e16. https://doi.org/10.1016/j.cell.2019.04.040
Becker AJ, McCULLOCH EA, Till JE (1963) Cytological demonstration of the clonal nature of spleen colonies derived from transplanted mouse marrow cells. Nature 197:452–454. https://doi.org/10.1038/197452a0
Beerman I, Maloney WJ, Weissmann IL, Rossi DJ (2010) Stem cells and the aging hematopoietic system. Curr Opin Immunol 22:500–506. https://doi.org/10.1016/j.coi.2010.06.007
Beerman I, Rossi DJ (2015) Epigenetic control of stem cell potential during homeostasis, aging, and disease. Cell Stem Cell 16:613–625. https://doi.org/10.1016/j.stem.2015.05.009
Bianco P (2014) "Mesenchymal" stem cells. Annu Rev Cell Dev Biol 30:677–704. https://doi.org/10.1146/annurev-cellbio-100913-013132
Bianco P, Cao X, Frenette PS et al (2013) The meaning, the sense and the significance: translating the science of mesenchymal stem cells into medicine. Nat Med 19:35–42. https://doi.org/10.1038/nm.3028
Bianco P, Robey PG (2015) Skeletal Stem Cells. Development 142:1023–1027. https://doi.org/10.1242/dev.102210
Bidegain M, Roos BA, Hill EL et al (1995) Calcitonin gene-related peptide (CGRP) in the developing mouse limb. Endocr Res 21:743–755. https://doi.org/10.1080/07435809509030488
Blank U, Karlsson S (2015) TGF-β signaling in the control of hematopoietic stem cells. Blood 125:3542–3550. https://doi.org/10.1182/blood-2014-12-618090
Boettcher S, Gerosa RC, Radpour R et al (2014) Endothelial cells translate pathogen signals into G-CSF-driven emergency granulopoiesis. Blood 124:1393–1403. https://doi.org/10.1182/blood-2014-04-570762
Bourgine PE, Martin I, Schroeder T (2018) Engineering human bone marrow proxies. Cell Stem Cell 22:298–301. https://doi.org/10.1016/j.stem.2018.01.002
Bourgine PE, Scotti C, Pigeot S et al (2014) Osteoinductivity of engineered cartilaginous templates devitalized by inducible apoptosis. PNAS 111:17426–17431. https://doi.org/10.1073/pnas.1411975111
Broxmeyer HE, Hoggatt J, O'Leary HA et al (2012) Dipeptidylpeptidase 4 negatively regulates colony-stimulating factor activity and stress hematopoiesis. Nat Med 18:1786–1796. https://doi.org/10.1038/nm.2991
Bruns I, Lucas D, Pinho S et al (2014) Megakaryocytes regulate hematopoietic stem cell quiescence through CXCL4 secretion. Nat Med 20:1315–1320. https://doi.org/10.1038/nm.3707
Calvi LM, Adams GB, Weibrecht KW et al (2003) Osteoblastic cells regulate the haematopoietic stem cell niche. Nature 425:841–846. https://doi.org/10.1038/nature02040

Cao Y-R, Shao Z-H, Liu H et al (2005) The response of bone marrow hematopoietic cells to G-CSF in paroxysmal nocturnal hemoglobinuria patients. Zhonghua Xue Ye Xue Za Zhi 26:235–238
Caplan AI (1991) Mesenchymal stem cells. J Orthop Res 9:641–650. https://doi.org/10.1002/jor.1100090504
Caplan AI (2008) All MSCs are pericytes? Cell Stem Cell 3:229–230. https://doi.org/10.1016/j.stem.2008.08.008
Caplan AI (2017) Mesenchymal stem cells: time to change the name! Stem Cells Transl Med 6:1445–1451. https://doi.org/10.1002/sctm.17-0051
Castro-Malaspina H, Gay RE, Resnick G et al (1980) Characterization of human bone marrow fibroblast colony-forming cells (CFU-F) and their progeny. Blood 56:289–301
Cawthorn WP, Scheller EL, Learman BS et al (2014) Bone marrow adipose tissue is an endocrine organ that contributes to increased circulating adiponectin during caloric restriction. Cell Metab 20:368–375. https://doi.org/10.1016/j.cmet.2014.06.003
Challen GA, Boles NC, Chambers SM, Goodell MA (2010) Distinct hematopoietic stem cell subtypes are differentially regulated by TGF-beta1. Cell Stem Cell 6:265–278. https://doi.org/10.1016/j.stem.2010.02.002
Chan CKF, Chen C-C, Luppen CA et al (2009) Endochondral ossification is required for haematopoietic stem-cell niche formation. Nature 457:490–494. https://doi.org/10.1038/nature07547
Chan CKF, Gulati GS, Sinha R et al (2018) Identification of the human skeletal stem cell. Cell 175:43–56.e21. https://doi.org/10.1016/j.cell.2018.07.029
Chan CKF, Lindau P, Jiang W et al (2013) Clonal precursor of bone, cartilage, and hematopoietic niche stromal cells. Proc Natl Acad Sci U S A 110:12643–12648. https://doi.org/10.1073/pnas.1310212110
Chan CKF, Seo EY, Chen JY et al (2015) Identification and specification of the mouse skeletal stem cell. Cell 160:285–298. https://doi.org/10.1016/j.cell.2014.12.002
Christensen JL, Wright DE, Wagers AJ, Weissman IL (2004) Circulation and chemotaxis of fetal hematopoietic stem cells. PLoS Biol 2:E75. https://doi.org/10.1371/journal.pbio.0020075
Christodoulou C, Spencer JA, Yeh S-CA et al (2020) Live-animal imaging of native haematopoietic stem and progenitor cells. Nature 578:278–283. https://doi.org/10.1038/s41586-020-1971-z
Corselli M, Chen C-W, Sun B et al (2012) The tunica adventitia of human arteries and veins as a source of mesenchymal stem cells. Stem Cells Dev 21:1299–1308. https://doi.org/10.1089/scd.2011.0200
Coşkun S, Chao H, Vasavada H et al (2014) Development of the fetal bone marrow niche and regulation of HSC quiescence and homing ability by emerging osteolineage cells. Cell Rep 9:581–590. https://doi.org/10.1016/j.celrep.2014.09.013
Crisan M, Yap S, Casteilla L et al (2008) A perivascular origin for mesenchymal stem cells in multiple human organs. Cell Stem Cell 3:301–313. https://doi.org/10.1016/j.stem.2008.07.003
Czechowicz A, Kraft D, Weissman IL, Bhattacharya D (2007) Efficient transplantation via antibody-based clearance of hematopoietic stem cell niches. Science 318:1296–1299. https://doi.org/10.1126/science.1149726
Debnath S, Yallowitz AR, McCormick J et al (2018) Discovery of a periosteal stem cell mediating intramembranous bone formation. Nature 562:133–139. https://doi.org/10.1038/s41586-018-0554-8
Demontiero O, Vidal C, Duque G (2012) Aging and bone loss: new insights for the clinician. Ther Adv Musculoskelet Dis 4:61–76. https://doi.org/10.1177/1759720X11430858
Ding L, Morrison SJ (2013) Haematopoietic stem cells and early lymphoid progenitors occupy distinct bone marrow niches. Nature 495:231–235. https://doi.org/10.1038/nature11885
Ding L, Saunders TL, Enikolopov G, Morrison SJ (2012) Endothelial and perivascular cells maintain haematopoietic stem cells. Nature 481:457–462. https://doi.org/10.1038/nature10783
Dominici M, Le Blanc K, Mueller I et al (2006) Minimal criteria for defining multipotent mesenchymal stromal cells. The international society for cellular therapy position statement. Cytotherapy 8:315–317. https://doi.org/10.1080/14653240600855905

Doro DH, Grigoriadis AE, Liu KJ (2017) Calvarial suture-derived stem cells and their contribution to cranial bone repair. Front Physiol 8. https://doi.org/10.3389/fphys.2017.00956
Dose K (1981) Ernst Haeckel's concept of an evolutionary origin of life. Biosystems 13:253–258. https://doi.org/10.1016/0303-2647(81)90005-8
Duchamp de Lageneste O, Julien A, Abou-Khalil R et al (2018) Periosteum contains skeletal stem cells with high bone regenerative potential controlled by Periostin. Nat Commun 9:773. https://doi.org/10.1038/s41467-018-03124-z
Emami A, Talaei-Khozani T, Tavanafar S et al (2020) Synergic effects of decellularized bone matrix, hydroxyapatite, and extracellular vesicles on repairing of the rabbit mandibular bone defect model. J Transl Med 18:361. https://doi.org/10.1186/s12967-020-02525-3
Essers MAG, Offner S, Blanco-Bose WE et al (2009) IFNalpha activates dormant haematopoietic stem cells in vivo. Nature 458:904–908. https://doi.org/10.1038/nature07815
Evans MJ, Kaufman MH (1981) Establishment in culture of pluripotential cells from mouse embryos. Nature 292:154–156. https://doi.org/10.1038/292154a0
Fan Y, Hanai J, Le PT et al (2017) Parathyroid hormone directs bone marrow mesenchymal cell fate. Cell Metab 25:661–672. https://doi.org/10.1016/j.cmet.2017.01.001
Farr JN, Xu M, Weivoda MM et al (2017) Targeting cellular senescence prevents age-related bone loss in mice. Nat Med 23:1072–1079. https://doi.org/10.1038/nm.4385
Feins S, Kong W, Williams EF et al (2019) An introduction to chimeric antigen receptor (CAR) T-cell immunotherapy for human cancer. Am J Hematol 94:S3–S9. https://doi.org/10.1002/ajh.25418
Fidler IJ (2003) The pathogenesis of cancer metastasis: the "seed and soil" hypothesis revisited. Nat Rev Cancer 3:453–458. https://doi.org/10.1038/nrc1098
Fischer UM, Harting MT, Jimenez F et al (2009) Pulmonary passage is a major obstacle for intravenous stem cell delivery: the pulmonary first-pass effect. Stem Cells Dev 18:683–692. https://doi.org/10.1089/scd.2008.0253
Fozza C, Crobu V, Isoni MA, Dore F (2016) The immune landscape of myelodysplastic syndromes. Crit Rev Oncol Hematol 107:90–99. https://doi.org/10.1016/j.critrevonc.2016.08.016
Friedenstein AJ, Piatetzky-Shapiro II, Petrakova KV (1966) Osteogenesis in transplants of bone marrow cells. J Embryol Exp Morphol 16:381–390
Galipeau J, Sensébé L (2018) Mesenchymal stromal cells: clinical challenges and therapeutic opportunities. Cell Stem Cell 22:824–833. https://doi.org/10.1016/j.stem.2018.05.004
Gamboa L, Zamat AH, Kwong GA (2020) Synthetic Immunity by Remote Control. Theranostics 10:3652–3667. https://doi.org/10.7150/thno.41305
Gao X, Zhang D, Xu C et al (2020) Nociceptive nerves regulate haematopoietic stem cell mobilization. Nature 1–6. https://doi.org/10.1038/s41586-020-03057-y
George BM, Kao KS, Kwon H-S et al (2019) Antibody conditioning enables MHC-mismatched hematopoietic stem cell transplants and organ graft tolerance. Cell Stem Cell 25:185–192.e3. https://doi.org/10.1016/j.stem.2019.05.018
Greenbaum A, Hsu Y-MS, Day RB et al (2013) CXCL12 in early mesenchymal progenitors is required for haematopoietic stem-cell maintenance. Nature 495:227–230. https://doi.org/10.1038/nature11926
Gulati GS, Murphy MP, Marecic O et al (2018) Isolation and functional assessment of mouse skeletal stem cell lineage. Nat Protoc 13:1294–1309. https://doi.org/10.1038/nprot.2018.041
Hallett SA, Matsushita Y, Ono W et al (2021) Chondrocytes in the resting zone of the growth plate are maintained in a Wnt-inhibitory environment. eLife 10:e64513. https://doi.org/10.7554/eLife.64513
Herber CB, Krause WC, Wang L et al (2019) Estrogen signaling in arcuate Kiss1 neurons suppresses a sex-dependent female circuit promoting dense strong bones. Nat Commun 10:163. https://doi.org/10.1038/s41467-018-08046-4
Ho TT, Dellorusso PV, Verovskaya EV et al (2021) Aged hematopoietic stem cells are refractory to bloodborne systemic rejuvenation interventions. J Exp Med 218:e20210223. https://doi.org/10.1084/jem.20210223

Ho Y-H, del Toro R, Rivera-Torres J et al (2019) Remodeling of bone marrow hematopoietic stem cell niches promotes myeloid cell expansion during premature or physiological aging. Cell Stem Cell 25:407–418.e6. https://doi.org/10.1016/j.stem.2019.06.007

Itkin T, Gur-Cohen S, Spencer JA et al (2016) Distinct bone marrow blood vessels differentially regulate haematopoiesis. Nature 532:323–328. https://doi.org/10.1038/nature17624

Jaiswal S, Ebert BL (2019) Clonal hematopoiesis in human aging and disease. Science 366. https://doi.org/10.1126/science.aan4673

Jaiswal S, Fontanillas P, Flannick J et al (2014) Age-related clonal hematopoiesis associated with adverse outcomes. N Engl J Med 371:2488–2498. https://doi.org/10.1056/NEJMoa1408617

James AW, Leucht P, Levi B et al (2010) Sonic Hedgehog influences the balance of osteogenesis and adipogenesis in mouse adipose-derived stromal cells. Tissue Eng Part A 16:2605–2616. https://doi.org/10.1089/ten.TEA.2010.0048

Josephson AM, Bradaschia-Correa V, Lee S et al (2019) Age-related inflammation triggers skeletal stem/progenitor cell dysfunction. PNAS 116:6995–7004. https://doi.org/10.1073/pnas.1810692116

Jovčevska I, Muyldermans S (2020) The therapeutic potential of nanobodies. BioDrugs 34:11–26. https://doi.org/10.1007/s40259-019-00392-z

Katayama Y, Battista M, Kao W-M et al (2006) Signals from the sympathetic nervous system regulate hematopoietic stem cell egress from bone marrow. Cell 124:407–421. https://doi.org/10.1016/j.cell.2005.10.041

Kenswil KJG, Pisterzi P, Sánchez-Duffhues G et al (2021) Endothelium-derived stromal cells contribute to hematopoietic bone marrow niche formation. Cell Stem Cell 28:65–3670. e11. https://doi.org/10.1016/j.stem.2021.01.006

Kfoury Y, Scadden DT (2015) Mesenchymal cell contributions to the stem cell niche. Cell Stem Cell 16:239–253. https://doi.org/10.1016/j.stem.2015.02.019

Kiel MJ, Yilmaz ÖH, Iwashita T et al (2005) SLAM family receptors distinguish hematopoietic stem and progenitor cells and reveal endothelial niches for stem cells. Cell 121:1109–1121. https://doi.org/10.1016/j.cell.2005.05.026

Kim YS, Majid M, Melchiorri AJ, Mikos AG (2018) Applications of decellularized extracellular matrix in bone and cartilage tissue engineering. Bioeng Transl Med 4:83–95. https://doi.org/10.1002/btm2.10110

Klichinsky M, Ruella M, Shestova O et al (2020) Human chimeric antigen receptor macrophages for cancer immunotherapy. Nat Biotechnol 38:947–953. https://doi.org/10.1038/s41587-020-0462-y

Koide Y, Morikawa S, Mabuchi Y et al (2007) Two distinct stem cell lineages in murine bone marrow. Stem Cells 25:1213–1221. https://doi.org/10.1634/stemcells.2006-0325

Kokkaliaris KD, Kunz L, Cabezas-Wallscheid N et al (2020) Adult blood stem cell localization reflects the abundance of reported bone marrow niche cell types and their combinations. Blood 136:2296–2307. https://doi.org/10.1182/blood.2020006574

Kunisaki Y, Bruns I, Scheiermann C et al (2013) Arteriolar niches maintain haematopoietic stem cell quiescence. Nature 502:637–643. https://doi.org/10.1038/nature12612

Kuribayashi W, Iwama A, Oshima M (2019) Incomplete rejuvenation of aged HSCs in young bone marrow niche. Exp Hematol 76:S72. https://doi.org/10.1016/j.exphem.2019.06.381

Kusumbe AP, Ramasamy SK, Itkin T et al (2016) Age-dependent modulation of vascular niches for haematopoietic stem cells. Nature 532:380–384. https://doi.org/10.1038/nature17638

Kuznetsov SA, Krebsbach PH, Satomura K et al (1997) Single-colony derived strains of human marrow stromal fibroblasts form bone after transplantation in vivo. J Bone Miner Res 12:1335–1347. https://doi.org/10.1359/jbmr.1997.12.9.1335

Lai W, Li Y, Mak S et al (2013) Reconstitution of bone-like matrix in osteogenically differentiated mesenchymal stem cell-collagen constructs: a three-dimensional in vitro model to study hematopoietic stem cell niche. J Tissue Eng 4:2041731413508668. https://doi.org/10.1177/2041731413508668

Lampreia FP, Carmelo JG, Anjos-Afonso F (2017) Notch signaling in the regulation of hematopoietic stem cell. Curr Stem Cell Rep 3:202–209. https://doi.org/10.1007/s40778-017-0090-8

Laurenti E, Göttgens B (2018) From haematopoietic stem cells to complex differentiation landscapes. Nature 553:418–426. https://doi.org/10.1038/nature25022

Leimkuehler NB, Gleitz HFE, Ronghui L et al (2021) Heterogeneous bone-marrow stromal progenitors drive myelofibrosis via a druggable alarmin axis. Cell Stem Cell 28:637–652. e8. https://doi.org/10.1016/j.stem.2020.11.004

Li D, Xue W, Li M et al (2018) VCAM-1+ macrophages guide the homing of HSPCs to a vascular niche. Nature 564:119–124. https://doi.org/10.1038/s41586-018-0709-7

Li J, Chen G, Xu X et al (2019) Advances of injectable hydrogel-based scaffolds for cartilage regeneration. Regenerative Biomaterials 6:129–140. https://doi.org/10.1093/rb/rbz022

Logan M, Martin JF, Nagy A et al (2002) Expression of Cre Recombinase in the developing mouse limb bud driven by a Prxl enhancer. Genesis 33:77–80. https://doi.org/10.1002/gene.10092

Mabuchi Y, Houlihan DD, Akazawa C et al (2013a) Prospective isolation of murine and human bone marrow mesenchymal stem cells based on surface markers. Stem Cells Int 2013. https://doi.org/10.1155/2013/507301

Mabuchi Y, Morikawa S, Harada S et al (2013b) LNGFR+THY-1+VCAM-1hi+ cells reveal functionally distinct subpopulations in mesenchymal stem cells. Stem Cell Reports 1:152–165. https://doi.org/10.1016/j.stemcr.2013.06.001

Marecic O, Tevlin R, McArdle A et al (2015) Identification and characterization of an injury-induced skeletal progenitor. Proc Natl Acad Sci USA 112:9920–9925. https://doi.org/10.1073/pnas.1513066112

Maruyama T, Jeong J, Sheu T-J, Hsu W (2016) Stem cells of the suture mesenchyme in craniofacial bone development, repair and regeneration. Nat Commun 7:10526. https://doi.org/10.1038/ncomms10526

Matatall KA, Shen C-C, Challen GA, King KY (2014) Type II interferon promotes differentiation of myeloid-biased hematopoietic stem cells. Stem Cells 32:3023–3030. https://doi.org/10.1002/stem.1799

Matthews BG, Novak S, Sbrana FV et al (2021) Heterogeneity of murine periosteum progenitors involved in fracture healing. elife 10:e58534. https://doi.org/10.7554/eLife.58534

McClune BL, Majhail NS (2013) Osteoporosis after stem cell transplantation. Curr Osteoporos Rep 11:305–310. https://doi.org/10.1007/s11914-013-0180-1

McMahon C, Baier AS, Pascolutti R et al (2018) Yeast surface display platform for rapid discovery of conformationally selective nanobodies. Nat Struct Mol Biol 25:289–296. https://doi.org/10.1038/s41594-018-0028-6

Mead AJ, Neo WH, Barkas N et al (2017) Niche-mediated depletion of the normal hematopoietic stem cell reservoir by Flt3-ITD-induced myeloproliferation. J Exp Med 214:2005–2021. https://doi.org/10.1084/jem.20161418

Medyouf H, Mossner M, Jann J-C et al (2014) Myelodysplastic cells in patients reprogram mesenchymal stromal cells to establish a transplantable stem cell niche disease unit. Cell Stem Cell 14:824–837. https://doi.org/10.1016/j.stem.2014.02.014

Méndez-Ferrer S, Michurina TV, Ferraro F et al (2010) Mesenchymal and haematopoietic stem cells form a unique bone marrow niche. Nature 466:829–834. https://doi.org/10.1038/nature09262

Mitroulis I, Kalafati L, Hajishengallis G, Chavakis T (2018) Myelopoiesis in the context of innate immunity. J Innate Immun 10:365–372. https://doi.org/10.1159/000489406

Mizoguchi T, Pinho S, Ahmed J et al (2014) Osterix marks distinct waves of primitive and definitive stromal progenitors during bone marrow development. Dev Cell 29:340–349. https://doi.org/10.1016/j.devcel.2014.03.013

Mizuhashi K, Ono W, Matsushita Y et al (2018) Resting zone of the growth plate harbors a unique class of skeletal stem cells. Nature 563:254–258. https://doi.org/10.1038/s41586-018-0662-5

Morikawa S, Mabuchi Y, Kubota Y et al (2009) Prospective identification, isolation, and systemic transplantation of multipotent mesenchymal stem cells in murine bone marrow. J Exp Med 206:2483–2496. https://doi.org/10.1084/jem.20091046

Morrison SJ, Hemmati HD, Wandycz AM, Weissman IL (1995) The purification and characterization of fetal liver hematopoietic stem cells. Proc Natl Acad Sci U S A 92:10302–10306. https://doi.org/10.1073/pnas.92.22.10302

Morrison SJ, Wright DE, Cheshier SH, Weissman IL (1997) Hematopoietic stem cells: challenges to expectations. Curr Opin Immunol 9:216–221. https://doi.org/10.1016/s0952-7915(97)80138-0

Murphy MP, Koepke LS, Lopez MT et al (2020) Articular cartilage regeneration by activated skeletal stem cells. Nat Med 1–10. https://doi.org/10.1038/s41591-020-1013-2

Murry CE, Keller G (2008) Differentiation of embryonic stem cells to clinically relevant populations: lessons from embryonic development. Cell 132:661–680. https://doi.org/10.1016/j.cell.2008.02.008

Nakamura-Ishizu A, Takubo K, Fujioka M, Suda T (2014) Megakaryocytes are essential for HSC quiescence through the production of thrombopoietin. Biochem Biophys Res Commun 454:353–357. https://doi.org/10.1016/j.bbrc.2014.10.095

Nakashima T, Hayashi M, Fukunaga T et al (2011) Evidence for osteocyte regulation of bone homeostasis through RANKL expression. Nat Med 17:1231–1234. https://doi.org/10.1038/nm.2452

Naveiras O, Nardi V, Wenzel PL et al (2009) Bone marrow adipocytes as negative regulators of the hematopoietic microenvironment. Nature 460:259–263. https://doi.org/10.1038/nature08099

Newton PT, Li L, Zhou B et al (2019) A radical switch in clonality reveals a stem cell niche in the epiphyseal growth plate. Nature 567:234–238. https://doi.org/10.1038/s41586-019-0989-6

Olsen BR, Reginato AM, Wang W (2000) Bone development. Annu Rev Cell Dev Biol 16:191–220. https://doi.org/10.1146/annurev.cellbio.16.1.191

Ortinau LC, Wang H, Lei K et al (2019) Identification of functionally distinct Mx1+αSMA+ periosteal skeletal stem cells. Cell Stem Cell 25:784-796.e5. https://doi.org/10.1016/j.stem.2019.11.003

Palchaudhuri R, Saez B, Hoggatt J et al (2016) Non-genotoxic conditioning for hematopoietic stem cell transplantation using a hematopoietic-cell-specific internalizing immunotoxin. Nat Biotechnol 34:738–745. https://doi.org/10.1038/nbt.3584

Pang WW, Price EA, Sahoo D et al (2011) Human bone marrow hematopoietic stem cells are increased in frequency and myeloid-biased with age. Proc Natl Acad Sci USA 108:20012–20017. https://doi.org/10.1073/pnas.1116110108

Pappenheim A (1896) Ueber Entwickelung und Ausbildung der Erythroblasten. Archiv f Pathol Anat 145:587–643. https://doi.org/10.1007/BF01969901

Park D, Spencer JA, Koh BI et al (2012) Endogenous bone marrow MSCs are dynamic, fate-restricted participants in bone maintenance and regeneration. Cell Stem Cell 10:259–272. https://doi.org/10.1016/j.stem.2012.02.003

Pasold J, Engelmann R, Keller J et al (2013) High bone mass in the STR/ort mouse results from increased bone formation and impaired bone resorption and is associated with extramedullary hematopoiesis. J Bone Miner Metab 31:71–81. https://doi.org/10.1007/s00774-012-0394-9

Pearson AC, Schiff M, Mrosek D et al (1986) Left ventricular diastolic function in weight lifters. Am J Cardiol 58:1254–1259. https://doi.org/10.1016/0002-9149(86)90392-9

Pietras EM, Mirantes-Barbeito C, Fong S et al (2016) Chronic interleukin-1 exposure drives haematopoietic stem cells towards precocious myeloid differentiation at the expense of self-renewal. Nat Cell Biol 18:607–618. https://doi.org/10.1038/ncb3346

Ping Z, Chen S, Hermans SJF et al (2019) Activation of NF-κB driven inflammatory programs in mesenchymal elements attenuates hematopoiesis in low-risk myelodysplastic syndromes. Leukemia 33:536–541. https://doi.org/10.1038/s41375-018-0267-x

Pinho S, Frenette PS (2019) Haematopoietic stem cell activity and interactions with the niche. Nat Rev Mol Cell Biol 20:303–320. https://doi.org/10.1038/s41580-019-0103-9

Pittenger MF, Mackay AM, Beck SC et al (1999) Multilineage potential of adult human mesenchymal stem cells. Science 284:143–147. https://doi.org/10.1126/science.284.5411.143

Pundole XN, Barbo AG, Lin H et al (2015) Increased incidence of fractures in recipients of hematopoietic stem-cell transplantation. JCO 33:1364–1370. https://doi.org/10.1200/JCO.2014.57.8195

Quéré R, Saint-Paul L, Carmignac V et al (2014) Tif1γ regulates the TGF-β1 receptor and promotes physiological aging of hematopoietic stem cells. Proc Natl Acad Sci U S A 111:10592–10597. https://doi.org/10.1073/pnas.1405546111

Raaijmakers MHGP, Mukherjee S, Guo S et al (2010) Bone progenitor dysfunction induces myelodysplasia and secondary leukemia. Nature 464:852–857. https://doi.org/10.1038/nature08851

Rajam M, Pulavendran S, Rose C, Mandal AB (2011) Chitosan nanoparticles as a dual growth factor delivery system for tissue engineering applications. Int J Pharm 410:145–152. https://doi.org/10.1016/j.ijpharm.2011.02.065

Ramasamy SK, Kusumbe AP, Itkin T et al (2016) Regulation of hematopoiesis and osteogenesis by blood vessel-derived signals. Annu Rev Cell Dev Biol 32:649–675. https://doi.org/10.1146/annurev-cellbio-111315-124936

Ransom RC, Carter AC, Salhotra A et al (2018) Mechanoresponsive stem cells acquire neural crest fate in jaw regeneration. Nature 563:514–521. https://doi.org/10.1038/s41586-018-0650-9

Reinisch A, Etchart N, Thomas D et al (2015) Epigenetic and in vivo comparison of diverse MSC sources reveals an endochondral signature for human hematopoietic niche formation. Blood 125:249–260. https://doi.org/10.1182/blood-2014-04-572255

Reinisch A, Hernandez DC, Schallmoser K, Majeti R (2017) Generation and use of a humanized bone-marrow-ossicle niche for hematopoietic xenotransplantation into mice. Nat Protoc 12:2169–2188. https://doi.org/10.1038/nprot.2017.088

Rossi DJ, Bryder D, Zahn JM et al (2005) Cell intrinsic alterations underlie hematopoietic stem cell aging. Proc Natl Acad Sci USA 102:9194–9199. https://doi.org/10.1073/pnas.0503280102

Rowe RG, Mandelbaum J, Zon LI, Daley GQ (2016) Engineering hematopoietic stem cells: lessons from development. Cell Stem Cell 18:707–720. https://doi.org/10.1016/j.stem.2016.05.016

Sacchetti B, Funari A, Michienzi S et al (2007) Self-renewing osteoprogenitors in bone marrow sinusoids can organize a hematopoietic microenvironment. Cell 131:324–336. https://doi.org/10.1016/j.cell.2007.08.025

Sacchetti B, Funari A, Remoli C et al (2016) No identical "mesenchymal stem cells" at different times and sites: human committed progenitors of distinct origin and differentiation potential are incorporated as adventitial cells in microvessels. Stem Cell Reports 6:897–913. https://doi.org/10.1016/j.stemcr.2016.05.011

Salazar VS, Gamer LW, Rosen V (2016) BMP signalling in skeletal development, disease and repair. Nat Rev Endocrinol 12:203–221. https://doi.org/10.1038/nrendo.2016.12

Saravanan S, Leena RS, Selvamurugan N (2016) Chitosan based biocomposite scaffolds for bone tissue engineering. Int J Biol Macromol 93:1354–1365. https://doi.org/10.1016/j.ijbiomac.2016.01.112

Scheller EL, Doucette CR, Learman BS et al (2015) Region-specific variation in the properties of skeletal adipocytes reveals regulated and constitutive marrow adipose tissues. Nat Commun 6:7808. https://doi.org/10.1038/ncomms8808

Schofield R (1978) The relationship between the spleen colony-forming cell and the haemopoietic stem cell. Blood Cells 4:7–25

Schwarz CS, Bucher CH, Schlundt C et al (2021) Spatio-temporal bone remodeling after hematopoietic stem cell transplantation. Int J Mol Sci 22:267. https://doi.org/10.3390/ijms22010267

Semerad CL, Christopher MJ, Liu F et al (2005) G-CSF potently inhibits osteoblast activity and CXCL12 mRNA expression in the bone marrow. Blood 106:3020–3027. https://doi.org/10.1182/blood-2004-01-0272

Shen B, Tasdogan A, Ubellacker JM et al (2021) A mechanosensitive peri-arteriolar niche for osteogenesis and lymphopoiesis. Nature 591:438–444. https://doi.org/10.1038/s41586-021-03298-5

Shiozawa Y, Pedersen EA, Havens AM et al (2011) Human prostate cancer metastases target the hematopoietic stem cell niche to establish footholds in mouse bone marrow. J Clin Invest 121:1298–1312. https://doi.org/10.1172/JCI43414

Shu HS, Liu YL, Tang XT et al (2021) Tracing the skeletal progenitor transition during postnatal bone formation. Cell Stem Cell. https://doi.org/10.1016/j.stem.2021.08.010

Sieber S, Wirth L, Cavak N et al (2018) Bone marrow-on-a-chip: long-term culture of human haematopoietic stem cells in a three-dimensional microfluidic environment. J Tissue Eng Regen Med 12:479–489. https://doi.org/10.1002/term.2507

Simmons PJ, Torok-Storb B (1991) Identification of stromal cell precursors in human bone marrow by a novel monoclonal antibody, STRO-1. Blood 78:55–62

Sipkins DA, Wei X, Wu JW et al (2005) In vivo imaging of specialized bone marrow endothelial microdomains for tumour engraftment. Nature 435:969–973. https://doi.org/10.1038/nature03703

Sipp D, Robey PG, Turner L (2018) Clear up this stem-cell mess. Nature 561:455–457. https://doi.org/10.1038/d41586-018-06756-9

Sisask G, Silfverswärd C-J, Bjurholm A, Nilsson O (2013) Ontogeny of sensory and autonomic nerves in the developing mouse skeleton. Auton Neurosci 177:237–243. https://doi.org/10.1016/j.autneu.2013.05.005

Spangrude GJ, Heimfeld S, Weissman IL (1988) Purification and characterization of mouse hematopoietic stem cells. Science 241:58–62. https://doi.org/10.1126/science.2898810

Sugiyama T, Nagasawa T (2012) Bone marrow niches for hematopoietic stem cells and immune cells. Inflamm Allergy Drug Targets 11:201–206. https://doi.org/10.2174/187152812800392689

Sivaraj KK, Jeong H-W, Dharmalingam B et al (2021) Regional specialization and fate specification of bone stromal cells in skeletal development. Cell Rep 36:109352. https://doi.org/10.1016/j.celrep.2021.109352

Taichman RS, Emerson SG (1994) Human osteoblasts support hematopoiesis through the production of granulocyte colony-stimulating factor. J Exp Med 179:1677–1682. https://doi.org/10.1084/jem.179.5.1677

Tavassoli M, Crosby WH (1968) Transplantation of marrow to extramedullary sites. Science 161:54–56. https://doi.org/10.1126/science.161.3836.54

Tevlin R, Seo EY, Marecic O et al (2017) Pharmacological rescue of diabetic skeletal stem cell niches. Sci Transl Med 9. https://doi.org/10.1126/scitranslmed.aag2809

Thevenot PT, Nair AM, Shen J et al (2010) The effect of incorporation of SDF-1α into PLGA scaffolds on stem cell recruitment and the inflammatory response. Biomaterials 31:3997–4008. https://doi.org/10.1016/j.biomaterials.2010.01.144

Tikhonova AN, Dolgalev I, Hu H et al (2019) The bone marrow microenvironment at single-cell resolution. Nature 569:222–228. https://doi.org/10.1038/s41586-019-1104-8

Till JE, McCulloch EA (1961) A direct measurement of the radiation sensitivity of normal mouse bone marrow cells. Radiat Res 14:213–222. https://doi.org/10.2307/3570892

Tormin A, Li O, Brune JC et al (2011) CD146 expression on primary nonhematopoietic bone marrow stem cells is correlated with in situ localization. Blood 117:5067–5077. https://doi.org/10.1182/blood-2010-08-304287

Urist MR (1970) A morphogenetic matrix for differentiation of bone tissue. Calcif Tissue Res (Suppl):98–101. https://doi.org/10.1007/BF02152373

Ventura Ferreira MS, Bergmann C, Bodensiek I et al (2016) An engineered multicomponent bone marrow niche for the recapitulation of hematopoiesis at ectopic transplantation sites. J Hematol Oncol 9:4. https://doi.org/10.1186/s13045-016-0234-9

Via AG, Frizziero A, Oliva F (2012) Biological properties of mesenchymal stem cells from different sources. Muscles Ligaments Tendons J 2:154–162

Weissman IL, Shizuru JA (2008) The origins of the identification and isolation of hematopoietic stem cells, and their capability to induce donor-specific transplantation tolerance and treat autoimmune diseases. Blood 112:3543–3553. https://doi.org/10.1182/blood-2008-08-078220

Whitlock CA, Tidmarsh GF, Muller-Sieburg C, Weissman IL (1987) Bone marrow stromal cell lines with lymphopoietic activity express high levels of a pre-B neoplasia-associated molecule. Cell 48:1009–1021. https://doi.org/10.1016/0092-8674(87)90709-4

Wolock SL, Krishnan I, Tenen DE et al (2019) Mapping distinct bone marrow niche populations and their differentiation paths. Cell Rep 28:302–311.e5. https://doi.org/10.1016/j.celrep.2019.06.031

Worthley DL, Churchill M, Compton JT et al (2015) Gremlin 1 identifies a skeletal stem cell with bone, cartilage, and reticular stromal potential. Cell 160:269–284. https://doi.org/10.1016/j.cell.2014.11.042

Yamazaki S, Ema H, Karlsson G et al (2011) Nonmyelinating schwann cells maintain hematopoietic stem cell hibernation in the bone marrow niche. Cell 147:1146–1158. https://doi.org/10.1016/j.cell.2011.09.053

Yu W, Zhong L, Yao L et al (2020) Bone marrow adipogenic lineage precursors (MALPs) promote osteoclastogenesis in bone remodeling and pathologic bone loss. J Clin Invest. https://doi.org/10.1172/JCI140214

Yue R, Zhou BO, Shimada IS et al (2016) Leptin receptor promotes adipogenesis and reduces osteogenesis by regulating mesenchymal stromal cells in adult bone marrow. Cell Stem Cell 18:782–796. https://doi.org/10.1016/j.stem.2016.02.015

Zhang J, Niu C, Ye L et al (2003) Identification of the haematopoietic stem cell niche and control of the niche size. Nature 425:836–841. https://doi.org/10.1038/nature02041

Zhao H, Feng J, Ho T-V et al (2015) The suture provides a niche for mesenchymal stem cells of craniofacial bones. Nat Cell Biol 17:386–396. https://doi.org/10.1038/ncb3139

Zhao M, Perry JM, Marshall H et al (2014) Megakaryocytes maintain homeostatic quiescence and promote post-injury regeneration of hematopoietic stem cells. Nat Med 20:1321–1326. https://doi.org/10.1038/nm.3706

Zhong L, Yao L, Tower RJ, et al (2020) Single cell transcriptomics identifies a unique adipose lineage cell population that regulates bone marrow environment. eLife 9:e54695. https://doi.org/10.7554/eLife.54695

Zhou BO, Yu H, Yue R et al (2017) Bone marrow adipocytes promote the regeneration of stem cells and haematopoiesis by secreting SCF. Nat Cell Biol 19:891–903. https://doi.org/10.1038/ncb3570

Zhou BO, Yue R, Murphy MM et al (2014) Leptin-receptor-expressing mesenchymal stromal cells represent the main source of bone formed by adult bone marrow. Cell Stem Cell 15:154–168. https://doi.org/10.1016/j.stem.2014.06.008

Cellular Niches for Hematopoietic Stem Cells and Lympho-Hematopoiesis in Bone Marrow During Homeostasis and Blood Cancers

Yoshiki Omatsu, Kei Higaki, and Takashi Nagasawa

Abstract Most types of blood cells, including immune cells are generated from hematopoietic stem cells (HSCs) within bone marrow in the adult. Most HSCs are in contact with and require the special microenvironment known as a niche for their maintenance. It has been thought that HSC niches comprise various types of support cells that provide critical signals, including cytokines and extracellular matrix for HSC regulation. However, among these cells, several lines of evidence have demonstrated that the population of bone marrow-specific mesenchymal stem cells, termed CXC chemokine ligand 12 (CXCL12)-abundant reticular (CAR) cells, which overlap strongly with leptin receptor-expressing ($LepR^+$) cells, is the major cellular component of HSC niches. CAR/$LepR^+$ cells give rise to most adipocytes and osteoblasts in adult bone marrow and express much higher levels of HSC niche factors, including cytokines CXCL12 and stem cell factor (SCF), which are essential for HSC maintenance, and transcription factors Foxc1 and Ebf3, which are essential for the formation and maintenance of HSC niches than other types of cells. CAR/$LepR^+$ cells are present in human bone marrow, undergo fibrotic expansion, and have reduced expression of HSC niche factors in hematopoietic malignancies.

1 Introduction

Most types of blood cells, including immune cells are generated from hematopoietic stem cells (HSCs) within liver in the fetus and bone marrow in the adult. HSCs are rare cells and cannot survive and self-renew on its own. Thus, most HSCs are in contact with and require the special microenvironment known as a niche for their maintenance. It has been thought that HSC niches comprise various types of support cells (cellular niche) that provide critical signals, including cytokines and extracellular matrix (ECM) for HSC regulation, including survival, proliferation,

Y. Omatsu · K. Higaki · T. Nagasawa (✉)
Laboratory of Stem Cell Biology and Developmental Immunology, Graduate School of Frontier Biosciences and Graduate School of Medicine, WPI Immunology Frontier Research Center, Osaka University, Suita 565-0871, Osaka, Japan
e-mail: tnagasa@fbs.osaka-u.ac.jp

T. Nagasawa (ed.), *Bone Marrow Niche*, Current Topics in Microbiology and Immunology 434, https://doi.org/10.1007/978-3-030-86016-5_2

and differentiation. Thus, the identity of HSC cellular niches has been a subject of longstanding debate. The first candidate cell type creating niches for HSCs is the population of bone lining osteoblasts, which are the chief bone-making cells (Zhang et al. 2003; Calvi et al. 2003; Arai et al. 2004). However, several lines of evidence argue against the relevance of osteoblasts in HSC maintenance (Kiel et al. 2007a; Bowers et al. 2015; Yu et al. 2015) and have demonstrated that the population of mesenchymal stem cells, termed CXC chemokine ligand 12 (CXCL12)-abundant reticular (CAR) cells, which overlap strongly with leptin receptor (LepR)-expressing ($LepR^+$) cells, is the major cellular component of HSC niches (Sugiyama et al. 2006; Omatsu et al. 2010, 2014; Ding et al. 2012). We will review recent studies on the identity and function of candidates for cells, which constitute a niche for HSCs and lympho-hematopoiesis in bone marrow.

2 CXCL12-Abundant Reticular (CAR) Cells/leptin Receptor-Expressing ($LepR^+$) Cells

2.1 *Identification of CAR/$LepR^+$ Cells*

To identify cellular niches for HSCs, Sugiyama et al. and Ding et al. focused their analysis on bone marrow cells which produce cytokines essential for the maintenance of HSCs (Sugiyama et al. 2006; Ding et al. 2012). The chemokine CXCL12 was known to be essential for colonization of bone marrow by HSCs during ontogeny and B cell development (Nagasawa et al. 1996; Tachibana et al. 1998; Ara et al. 2003; Zou et al. 1998). It has been shown that the numbers of HSCs, common lymphoid progenitors (CLPs), which give rise to B cell, plasmacytoid dendric cells (pDCs), T cells, and NK cells (Karsunky et al. 2008), pro-B cells, pre-B cells, B cells, pDCs, and NK cells were severely reduced in mice, in which CXCR4, the primary receptor for a chemokine CXCL12 is deleted in adults (pIpC-treated Mx1-Cre; $CXCR4^{flox/flox}$ mice) (Sugiyama et al. 2006; Kohara et al. 2007; Noda et al. 2011). Greenbaum et al., showed that the numbers of HSCs and CLPs were severely reduced in Prx1-Cre; $CXCL12^{flox/flox}$ mice, in which CXCL12 is deleted from non-endothelial mesenchymal cells, including all CAR cells and osteoblasts, in developing limbs (Greenbaum et al. 2013). These results indicate that CXCL12-CXCR4 axis is essential for the maintenance of HSCs and development of immune cells, including B cells, pDCs, and NK cells.

Based on this, Sugiyama et al. generated mice, in which the green fluorescent protein (GFP) reporter genes was knocked into the CXCL12 locus (CXCL12-GFP knockin mice) and showed a population of reticular cells with long processes express much higher levels of CXCL12 than any other types of cells in bone marrow and other organs and termed these cells CAR cells (Sugiyama et al. 2006) (Fig. 1a). CAR cells are scattered throughout bone marrow and their number is markedly larger than that of HSCs (Sugiyama et al. 2006; Omatsu et al. 2010; Ding et al. 2012;

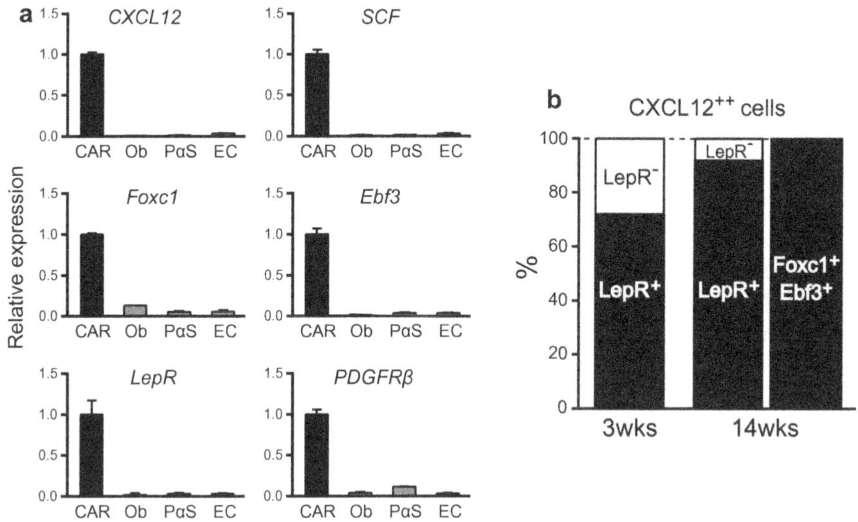

Fig. 1 CXCL12, SCF, Foxc1, Ebf3, and LepR are preferentially expressed in CAR/LepR$^+$ cells. **a** Expression levels of CXCL12, SCF, Foxc1, Ebf3, LepR, and PDGFRβ in CAR cells, osteoblasts (Ob), PαS mesenchymal progenitors, and endothelial cells (EC). **b** Frequencies of LepR-expressing cells in CAR cells at the age of 3 and 14 weeks

Shimoto et al. 2017). CAR cells were also isolated as CD45$^-$Ter119$^-$CD31$^-$Sca-1$^-$platelet-derived growth factor receptor (PDGFR)-β^{high} cells because PDGFR-β is preferentially expressed in CAR cells in the bone marrow (Omatsu et al. 2010, 2014).

Stem cell factor (SCF) was known to be essential for the maintenance of HSCs (Barker 1994; Miller et al. 1996) and erythroid precursors (Barker 1994) and is preferentially expressed in CAR cells (Omatsu et al. 2010; Ding et al. 2012) (Fig. 1a). Short-term ablation of CAR cells in vivo using mice, in which a transgene encoding the Diphtheria toxin (DT) receptor (DTR) was knocked into the CXCL12 locus (Omatsu et al. 2010) led to a marked reduction in production of the cytokine CXCL12 and SCF in bone marrow, indicating that CAR cells are the major producer of CXCL12 and SCF in the bone marrow (Omatsu et al. 2010).

Subsequently, Ding et al. generated mice, in which GFP reporter gene was knocked into the SCF locus (SCF-GFP knockin mice) and showed that cells that express much higher levels of SCF than any other types of bone marrow cells overlap strongly with CAR cells and most of these cells express much higher levels of LepR than any other types of marrow cells (Ding et al. 2012) (Fig. 1a). LepR is highly expressed in about 72 and 92% of CAR cells at the age of 3 and 14 weeks, respectively (Omatsu et al. 2014; Ding and Morrison 2013) (Fig. 1b). In addition, CAR/LepR$^+$ cells overlap strongly with Cre-expressing cells in NG2-Cre transgenic mice (Asada et al. 2017) or Nestin-GFP^{low} cells, which express lower levels of GFP than periarteriolar Nestin-GFP^{high} cells in a strain of transgenic mice, in which GFP is expressed under the control of the neural-specific regulatory elements of nestin gene (Zhou et al. 2014;

Méndez-Ferrer et al. 2010). However, endogenous Nestin or NG2 is not expressed in adult CAR/LepR$^+$ cells, or Nestin-GFP^{low} cells (Zhou et al. 2014; Kunisaki et al. 2013).

2.2 Anatomical Location of CAR/LepR$^+$ Cells and HSCs

It has been shown that CD150$^+$CD48$^-$CD41$^-$Lin$^-$ cells are enriched for long-term (LT) -HSCs and can be visualized in bone marrow sections (Kiel et al. 2005). Immunohistochemical analysis showed that about 97% of CD150$^+$CD48$^-$CD41$^-$ HSCs were scattered throughout bone marrow and were in contact with CAR/LepR$^+$ cells (Sugiyama et al. 2006). Acar et al. generated mice, having GFP reporter gene knocked into the locus of α-catulin, which is specifically expressed in HSCs, and imaged α-catulin-GFP$^+$c-kit$^+$ cells as HSCs (Acar et al. 2015). α-catulin-GFP$^+$c-kit$^+$ cells included almost all HSCs and about 30% of α-catulin-GFP$^+$c-kit$^+$ cells were LT-HSCs (Acar et al. 2015). Histological analysis showed that α-catulin-GFP$^+$c-kit$^+$ HSCs were more likely to be closed to CAR/LepR$^+$ cells than random spots and nearly all α-catulin-GFP$^+$c-kit$^+$ HSCs contact CAR/LepR$^+$ cells (Acar et al. 2015). Because H2B-GFP label retention has been shown to correlate with long-term repopulation potential, functional LT-HSCs are defined as H2B-GFP$^+$c-kit$^+$ cells, termed H2B label-retaining (LR)-HSCs, which could be visualized in TetOP-H2B-GFP mice 20 weeks after the doxycycline administration (Shimoto et al. 2017; Foudi et al. 2009). Histological analysis showed that most of H2B-GFP$^+$c-kit$^+$ HSCs as well as are significantly more close to CAR cells than in random location (Shimoto et al. 2017).

Shimoto et al. show that upon transplantation of purified HSCs into normal mice not exposed to myeloablation, most donor HSCs engrafted into bone marrow and these HSCs were found in contact with CAR/LepR$^+$ cells distant from CAR/LepR$^+$ cells in contact with endogenous HSCs (Shimoto et al. 2017). In addition, intravital imaging of HSCs using HSC-specific reporter mice with tamoxifen-inducible Cre recombinase expressed from the HSC-specific Pdzk1ip1 gene showed that HSCs moved in the perivascular space throughout bone marrow and form close transient contacts with CAR/LepR$^+$ cells throughout their movement (Upadhaya et al. 2020). These results support the concept that all CAR/LepR$^+$ cells create facultative niches for HSCs.

2.3 Roles of CAR/LepR$^+$ Cells in the Maintenance of HSCs and Lympho-Hematopoiesis

Short-term selective ablation of CAR cells in vivo using mice, in which a transgene encoding the DTR was knocked into the CXCL12 locus (Omatsu et al. 2014) or DTR is expressed in LepR$^+$ cells (Zhou et al. 2014), led to a marked reduction in the numbers of functional and phenotypic LT-HSCs, cycling LT-HSCs, CLPs, granulocyte and macrophage progenitors (GMPs), and megakaryocyte and erythroid progenitors (MEPs). When SCF was conditionally deleted from LepR$^+$ cells using mice, in which Cre recombinase was knocked into the LepR locus (LepR-Cre mice), the numbers of functional and phenotypic LT-HSCs (Ding et al. 2012), CLPs (Comazzetto et al. 2019), and MEPs (Comazzetto et al. 2019) were severely reduced in bone marrow. Gomes et al. showed that about 62% of CAR cells express interleukin (IL)-7 (Cordeiro Gomes et al. 2016), which is essential for B cell and T cell development. When IL-7 was conditionally deleted from LepR$^+$ cells using LepR-Cre mice, the numbers of pro-B, and pre-B cells were severely reduced in bone marrow (Cordeiro Gomes et al. 2016). These results strongly suggest that CAR/LepR$^+$ cells play major roles in maintaining HSCs and lympho-hematopoietic progenitors.

2.4 CAR/LepR$^+$ Cells Are Mesenchymal Stem Cells

Historically, the term mesenchymal stem cells (MSCs) was classically defined as the spindle-shaped cells that proliferate in vitro as plastic-adherent cells, form colonies in vitro as colony-forming unit fibroblasts (CFU-Fs), and can differentiate into osteoblasts, chondrocytes, and adipocytes both in vitro and after transfer (Friedenstein et al. 1974; Pittenger et al. 1999). More recently, however, the term stem cell is defined by the ability to replenish the stem cell pool throughout life (self-renewal), that allow them to sustain tissue maintenance in vivo as well as the ability to differentiate into multiple mature cell types (multipotency). Based on in vitro studies, the majority of CAR/LepR$^+$ cells have been shown to have potential to differentiate into adipocytes and osteoblasts (Omatsu et al. 2010). Zhou et al. show that LepR$^+$ cells are enrich for cells that form CFU-Fs and give rise to most adipocytes and osteoblasts but few chondrocytes in the bone marrow, using LepR-Cre; Rosa26-CAG-loxp-tdTomato transgenic mice (Zhou et al. 2014). Seike et al. performed lineage tracing where they transiently and specifically marked CAR/LepR$^+$ cells, in which the transcription factor Ebf3 is specifically expressed (Seike et al. 2018), using Ebf3-CreERT2; Rosa26-CAG-loxp-tdTomato transgenic mice and showed that Ebf3-expressing CAR cells give rise to most adipocytes and osteoblasts and remained stable over the course of 13 months (Seike et al. 2018). These results indicate that CAR/LepR$^+$ cells are self-renewing mesenchymal stem cells.

2.5 Molecular Basis for Formation and Maintenance of CAR/LepR⁺ Cells

It has been shown that transcription factors Foxc1 and Ebf3 are preferentially expressed in CAR cells (Fig. 1a) (Omatsu et al. 2014; Seike et al. 2018). Foxc1 protein belongs to a family of transcription factors characterized by the presence of a forkhead box (Fox) DNA-binding domain. Foxc1 is also expressed in arachnoid cells and in the mesenchymal layer above them in the developing meninges in the brain and Foxc1 null mutants die perinatally with hemorrhagic hydrocephalus and calvarial defects (Kume et al. 1998). Single cell RT-PCR analysis showed that all individual CAR cells expressed Foxc1 in 14-week-old mice (Omatsu et al. 2014). When Foxc1 was conditionally deleted from CAR/LepR$^+$ cells using LepR-Cre; Foxc1$^{flox/flox}$ mice, the numbers of LT-HSCs, MEPs, GMPs, CLPs, pro-B, and pre-B cells were severely reduced in bone marrow and the numbers of adipocytes were markedly increased in the bone marrow (Omatsu et al. 2014) (Fig. 2). Ebf3 protein belongs to the Ebf family, which have a unique DNA-binding domain in addition to a dimerization domain with homology to the classical basic helix-loop-helix (bHLH) proteins. Of note, the Ebf *Drosophila* counterpart Collier is expressed in candidate cellular niches for blood cells and essential for hematopoiesis (Crozatier et al. 2004). Ebf1, which is expressed in B cell precursors and critical for B cell development, is preferentially expressed in CAR cells among bone marrow non-hematopoietic cells (Seike et al. 2018). When both Ebf3 and Ebf1 was conditionally deleted from CAR/LepR$^+$ cells using LepR-Cre; Foxc1$^{flox/flox}$ mice, the numbers of LT-HSCs,

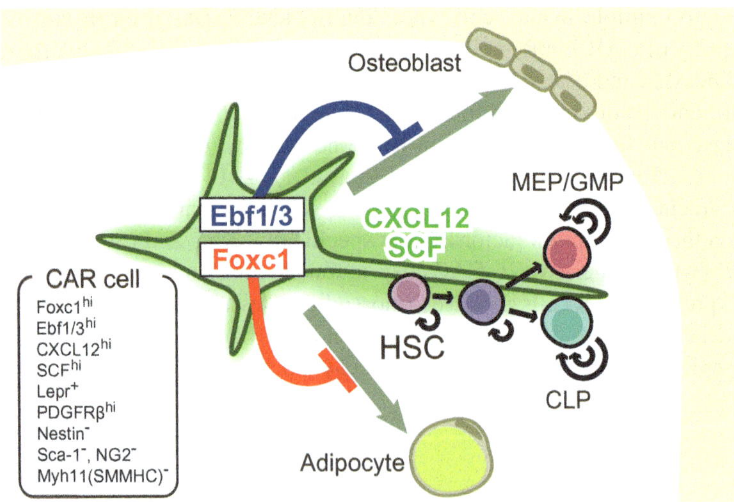

Fig. 2 Transcription factors Foxc1 and Ebf3 are essential for formation and maintenance of HSC niches. Foxc1 and Ebf1/3 enhance CXCL12 and SCF expression and play a critical role in maintaining HSPCs and inhibiting differentiation of CAR cells into adipocytes and osteoblasts

MEPs, GMPs, CLPs, pro-B, and pre-B cells were severely reduced in bone marrow and bone was markedly increased in the bone marrow (Seike et al. 2018) (Fig. 2). Experiments using Ebf1 knockout Ebf3 heterozygous and Ebf1 heterozygous Ebf3 knockout mice show that contribution of Ebf3 to bone marrow HSPC niches and cavity maintenance is greater than that of Ebf1 (Seike et al. 2018). It has been reported that enforced expression of five genes encoding transcription factors, including Klf7, Ostf1, Xbp1, Irf3, and Irf7, restored HSC niche function impaired in cultured bone marrow-derived MSC (Nakahara et al. 2019). Further studies will determine the role of these transcription factors in the formation and/or maintenance of HSC cellular niches in vivo.

2.6 The Bone Marrow Microenvironment at Single-Cell Resolution

Molecular definition of the bone marrow non-hematopoietic cell populations has been provided with single-cell RNA sequencing. Four groups identified a population that showed a high transcriptomic similarity to pre-defined CAR/LepR$^+$ cells (Tikhonova et al. 2019; Baryawno et al. 2019; Wolock et al. 2019; Baccin et al. 2020). Within CAR/LepR$^+$ cell population, Tikhonova et al. identified four subclusters, including matrix Gla protein (Mgp)^{high}, lipoprotein lipase (Lpl)^{high}, Wnt inhibitory factor 1 (Wif1)^{high}, and osteopontin (Spp1)^{high}bone sialoprotein (Ibsp)^{high} subclusters (Tikhonova et al. 2019). Baryawno et al. identified four subclusters within the CAR cell population distinguished by the expression levels of CXCL12, LepR, Grem1, and multiple other genes (Baryawno et al. 2019). Baccin et al. identified two subclusters within the CAR cell population and termed these subclusters Adipo-CAR and Osteo-CAR cells (Baccin et al. 2020). The Adipo-CAR cells comprise about 93% of CAR cells and express higher levels of LepR (Baccin et al. 2020). In contrast, Osteo-CAR cells comprise about 7% of CAR cells and express higher levels of Osterix and lower levels of LepR (Baccin et al. 2020). Wolock et al. mapped differentiation path of cells in CAR cell population to the osteoblasts, chondrocytes, and adipocytes (Wolock et al. 2019). Cells in all these subclusters might have potential to support HSCs and lympho-hematopoiesis since they express high levels of CXCL12, SCF, Foxc1, and Ebf3, which are essential for formation of HSPC niches (Baryawno et al. 2019; Baccin et al. 2020). However, it will be interesting to determine more precisely what niche function and behavior are specific for each subcluster. In addition, analyses of bone marrow cells using single-cell RNA sequencing have confirmed that endogenous Nestin or NG2 is not expressed in all subclusters of CAR/LepR$^+$ cells and have shown that endogenous Nestin is expressed in parts of arterial endothelial cells and that endogenous NG2 is expressed in chondrocytes and fibroblasts (Baryawno et al. 2019; Baccin et al. 2020).

2.7 Human CAR/LepR$^+$ Cells

In 1993, Cattoretti et al. showed that reticular cells with long processes express CD271 in human bone marrow (Cattoretti et al. 1993). Subsequently, human bone marrow CD271$^+$ non-hematopoietic cells have been shown to contain CFU-Fs and cells which have potential to differentiate into osteoblasts and adipocytes (Quirici et al. 2002; Jones et al. 2002). CD271$^+$ non-hematopoietic cells express higher levels of CXCL12 and LepR than hematopoietic cells (Churchman et al. 2012) and the PDGFR$\alpha^{lo/-}$ subpopulation of CD271$^+$ non-hematopoietic cells are enrich for CFU-Fs, have ability to support CD34$^+$ primitive hematopoietic cells in vitro, and express higher levels of CXCL12 and LepR compared with CD271$^+$PDGFRα^+ non-hematopoietic cells (Li et al. 2014). Histological analysis showed that most CD34$^+$ primitive hematopoietic cells were in contact with CD271$^+$ cells (Tormin et al. 2011; Flores-Figueroa et al. 2012) but not nestin$^+$ cells (Flores-Figueroa et al. 2012). On the other hand, it has been reported that CD146-expressing cells surrounding sinusoidal endothelial cells in human bone marrow express Angiopoietin-1 and can form hematopoietic microenvironments at heterotopic sites when transplanted subcutaneously into immunocompromised mice (Sacchetti et al. 2007). However, Tormin et al. showed that about 66% of CD146$^+$ cells are negative for CD271 and do not contain CFU-Fs.

Tormin et al. (2011). Together with the result that CD271$^+$CD146$^-$ cells are bone lining osteoblasts (Tormin et al. 2011), these results suggest that CD271$^+$CD146$^+$ or CD271$^+$PDGFR$\alpha^{lo/-}$ cells are mesenchymal progenitors which function as niches for HSPCs. Recently, Aoki et al. have shown that LepR$^+$ non-hematopoietic cells, which are positive for CD271 and contain both CD146$^+$ and CD146$^-$ cells, express much higher levels of HSC niche factors, including CXCL12, SCF, Foxc1, and Ebf3, and that CD271$^+$LepR$^-$PDGFR$\alpha^{lo/-}$ cells are osteoblasts (Aoki et al. 2021). This suggests that LepR$^+$ non-hematopoietic cells are the human counterpart of CAR/LepR$^+$ cell, which are characterized by salient features (Aoki et al. 2021).

2.8 CAR/LepR$^+$ Cells in Leukemia

Agarwal et al. show that deletion of CXCL12 from non-endothelial bone marrow non-hematopoietic cells, including CAR/LepR$^+$ cells, reduced normal HSCs but promoted expansion of leukemic stem cells (LSCs) by increasing self-renewing cell divisions, and increased LSC elimination by tyrosine kinase inhibitors (TKIs) in the mouse model of chronic myeloid leukemia (CML), using the well-characterized SCL-tTA/BCR-ABL transgenic inducible CML mouse model (Agarwal et al. 2019). These results suggest that CXCL12 maintains quiescence of TKI-resistent LSCs and inhibits LSC expansion in CML (Aoki et al. 2021). On the other hand, it has been shown that CXCL12 expression levels are reduced in bone marrow non-hematopoietic cells in the SCL-tTA/BCR-ABL transgenic CML mice and this reduction is only partially restored after treatment of TKIs (Zhang et al. 2012). Considering that CAR cells are the major producer of CXCL12 in the bone marrow, these results suggest that CXCL12 expression levels in CAR/LepR$^+$ cells are reduced in the CML mouse model, reducing normal HSCs but promoting expansion of LSCs in the CML mouse model. Consistent with this, expression of HSC niche factors, including Foxc1, Ebf3, CXCL12, and SCF was reduced in human CAR/LepR$^+$ cells sorted from bone marrow aspirates of CML patients (Aoki et al. 2021).

Furthermore, Decker et al. show that CAR/LepR$^+$ cells were the source of myofibroblasts and underwent expansion in primary myelofibrosis (PMF), using a PMF mouse model where thrombopoietin (TPO)-overexpressing retrovirally infected bone marrow cells were transplanted into irradiated mice (Decker et al. 2017).

2.9 Similarity Between CAR/LepR$^+$ Cells and Cellular Niches for Intestinal Stem Cells

Like hematopoietic cells, the intestinal epithelium is known to be the rapidly self-renewing tissue in adults. Intestinal epitherial stem (IES) cells give rise to all intestinal epithelial cells (Barker et al. 2007; Sangiorgi and Capecchi 2008) and have been thought to be in contact with and require the niche for their maintenance. Wnt signaling molecules are essential for the homeostasis of the intestinal epithelium (Korinek et al. 1998). Sato et al. reported that a cell type of terminally differentiated epithelial cells termed Paneth cells, which are known to secret bactericidal molecules, including defensins, produce Wnt signaling molecules and act as a cellular niche for IES cells in small intestine (Sato et al. 2011). However, homeostasis of the small intestine epithelium was unaltered in mice that lacked Paneth cells after conditional deletion of Math1 in intestinal epithelial cells (Kim et al. 2012; Durand et al. 2012) or mice, in which Wntless (Wls) a key protein required for the secretion of all Wnt proteins, was deleted from intestinal epithelial cells, including Paneth cells (Kabiri et al. 2014). These results indicate that Paneth cells are not essential for the maintenance of IES cells. In 2018, Shoshkes-Carmel et al. found that the population of reticular cells that express Foxl1 and form a subepithelial plexus that extend from the stomach to colon, termed Foxl1$^+$ telocytes, are the important source of Wnt proteins and essential for homeostasis of the intestinal epithelium (Shoshkes-Carmel et al. 2018). They showed that conditional deletion of porcupine (porcn), which is required for functional maturation of all Wnt proteins in Foxl1-expressing cells, resulted in severely decreased proliferation of IES cells and impaired intestinal epithelium (Shoshkes-Carmel et al. 2018) (Fig. 3).

Degirmenci et al. showed that conditional deletion of Wls in Gli1-expressing cells, including non-endothelial intestinal subepithelial mesenchymal cells, caused severely impaired colonic epithelium (Degirmenci et al. 2018). These results suggest

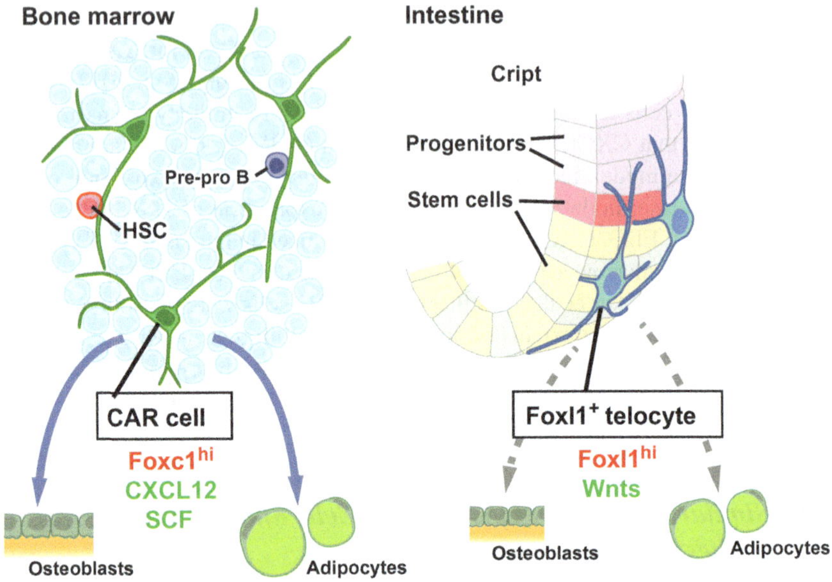

Fig. 3 CAR/LepR$^+$ cells share features with cellular niches for intestinal stem cells. Foxl1$^+$telocytes create niches for intestinal stem cells, have long processes, and contain mesenchymal progenitors that can differentiate into smooth muscle, osteoblasts, and adipocytes

that Gli1-expressing cells contain Foxl1$^+$ telocytes. Interestingly, like CAR/LepR$^+$ cells, Foxl1$^+$telocytes/Gli1$^+$ cells have long processes (Shoshkes-Carmel et al. 2018) and contained mesenchymal progenitors that can be differentiated in vitro into smooth muscle, osteoblasts, and adipocytes (Degirmenci et al. 2018). Thus, there exist similarities in properties between cellular niches for stem cells, CAR/LepR$^+$ cells and Foxl1$^+$ telocytes (Fig. 3).

3 Adipocytes

Although adipocytes in bone marrow are rare in young mouse, they increased during aging and after myeloablation. When in vitro culture system whereby proliferation of granulocyte precursor cells can be maintained in vitro for several months was described for the first time, giant fat cells in the adherent populations were speculated to be important for HSC maintenance (Dexter et al. 1977). However, Naveiras et al. showed that engraftment of hematopoietic cells after irradiation was accelerated in lipoatrophic A-ZIP/F1 fatless mice, which are genetically incapable of generating adipocytes, or mice treated with the perioxisome proliferator-activated receptor-g inhibitor, which inhibits adipocyte generation, compared with wild-type or untreated mice, suggesting that adipocytes are negative regulators of hematopoiesis

(Naveiras et al. 2009). In contrast, Zhou et al. showed that deletion of SCF from Adipoq-CreERT$^+$ adipocyte progenitors, which represent about 5% of CAR/LepR$^+$ cells, using Adipoq-CreERT; SCF$^{flox/flox}$ mice inhibited hematopoietic regeneration after irradiation or 5-fluorouracil treatment, suggesting that adipocytes promote hematopoietic regeneration (Zhou et al. 2017). Further studies will be needed to determine the role of adipocytes and their progenitors in regulation of HSCs and lympho-hematopoiesis.

4 Osteoblasts /N-cadherin$^+$CD45$^-$ Osteoblastic (SNO) Cells

4.1 Regulation of HSCs

In 2003, Calvi et al. showed that transgenic mice expressing constitutively active PTH/PTHrP receptor (PPR) under the control of the type I a1 collagen (Col1a1) promoter (Col 2.3 promoter) had increased numbers of trabecular bones and primitive hematopoietic cells and suggested that PPR-expressing osteoblasts were involved in regulation of HSCs (Calvi et al. 2003). Osteoblasts are lining the endosteal surface of bone and produce a unique combination of extracellular proteins, including osteocalcin, alkaline phosphatase and Col1a. On the other hand, Zhang et al. have defined BrdU-label-retaining cells after 10 days BrdU administration and 70 days chase as HSCs and revealed that these cells are in contact with a population of osteoblasts lining the bone surface, termed spindle-shaped N-cadherin$^+$CD45$^-$ osteoblastic (SNO) cells, which express a high level of N-cadherin (endosteal niche) (Zhang et al. 2003). They showed that mice, in which bone morphogenetic protein (BMP) receptor type IA (BMPRIA) and BMP receptor type IB (BMPRIB) are deleted using a polyI:C-inducible Mx1-Cre mouse line, had increased numbers of trabecular bones and primitive hematopoietic cells (Zhang et al. 2003). Based on these results, SNO cells were thought to function as a niche for HSCs. Arai et al. reported that Angiopoietin-1 was expressed in osteoblasts and supported quiescence of HSCs (Arai et al. 2004). Intravital imaging of HSCs using HSC-specific reporter mice with the GFP-myelodysplastic syndrome 1 (Mds1)-evi1 gene fusion product revealed that Mds1-evi1$^+$Flt3$^-$ HSCs were found close to the endosteum (Christodoulou et al. 2020).

However, it has been shown that most of BrdU-label-retaining cells after 10 days BrdU administration and 70 days chase are not included in the CD150$^+$CD48$^-$CD41$^-$Lin$^-$Sca-1$^+$c-kit$^+$ LT-HSC population (Kiel et al. 2007b). Immunohistochemical analysis have shown that the vast majority of CD150$^+$CD48$^-$CD41$^-$Lin$^-$ HSCs and α-catulin-GFP$^+$c-kit$^+$ HSCs are not in contact with the bone surface but scattered throughout the bone marrow cavity (Kiel et al. 2007a, 2005; Acar et al. 2015). On the other hand, specific deletion of osteoblasts in vivo using transgenic mice expressing herpesvirus thymidine kinase (TK) gene under the control of Col 2.3 promoter (Bowers et al. 2015) or the transcription factor Osterix promoter (Terashima et al. 2016) allowing the conditional ablation of TK-expressing cells after treatment with ganciclovir (GCV) did not decrease the numbers and functions of HSCs. In addition, mice in which Cre recombinase produced by the Osteocalcin (Ocn) promoter drives expression of the DTR on cell surface allowing the conditional ablation of DTR-expressing cells after treatment with DT did not alter the numbers and functions of HSCs (Yu et al. 2015). Furthermore, recent studies have shown that Angiopoietin-1 is not required for the maintenance of HSCs and hematopoiesis (Zhou et al. 2015). These results argue against the relevance of osteoblasts in the maintenance of HSCs. Further studies will be needed to determine the role of bone lining osteoblasts in HSC regulation.

4.2 Regulation of CLPs

It has been shown that mice, in which CXCL12 is deleted in cells expressing Col1a under the control of Col 2.3 promoter (Col 2.3-Cre; CXCL12$^{flox/flox}$ mice), have normal numbers of pro-B and pre-B cells but reduced numbers of CLPs (Ding et al. 2012). In addition, mice, in which IL-7 is deleted in cells expressing the transcription factor Osterix (Osterix-Cre; IL-7$^{flox/flox}$ mice), have reduced numbers of CLPs and pre-B cells (CFU-IL7) (Terashima et al. 2016), suggesting that osteoblasts are involved in B lymphopoiesis. In contrast, mice, in which IL-7 is deleted in cells expressing Col1a under the control of Col 2.3 promoter (Col 2.3-Cre;IL-7$^{flox/flox}$ mice), have normal numbers of CLPs and B cells and B cell precursors, including pro-B and pre-B cells (Cordeiro Gomes et al. 2016). Together with the results that Cre recombinase is expressed in the majority of CAR/LepR^{+} cells as well as all osteoblasts in Osterix-Cre- and Ocn-Cre-transgenic mice (Greenbaum et al. 2013; Zhang and Link 2016), this study suggests that osteoblast-derived IL-7 is not essential for B lymphopoiesis (Cordeiro Gomes et al. 2016).

5 Sinusoidal Endothelial Cells

In bone, most arteries of bone marrow enter the bone marrow cavities through the cortical bone and continue into typical capillaries with a complex system of thin-walled sinusoids which ramify throughout the marrow cavity (Bruyn et al. 1970). Some of these capillaries have an open lumen, and permits a slow blood flow and the passage of mature blood cells that are generated in bone marrow. Several studies have linked bone marrow endothelial cells to the control of HSC activity. Conditional deletion of VEGFR2 in adult mice, which inhibits regeneration of sinusoidal endothelial cells, prevents hematopoietic reconstitution in sublethally irradiated animals (Hooper et al. 2009). Kiel et al. have shown that about 60% of CD150^{+}CD48^{-}CD41^{-}Lin^{-}

HSCs are associated with sinusoidal endothelium (Kunisaki et al. 2013). Additionally, Ding et al. have shown that bone marrow endothelial cells express CXCL12 and SCF although their expression levels were much lower than those in CAR/LepR$^+$ cells and that number of CLPs and MEPs was unaltered but HSC number was reduced by about twofold when CXCL12 (Ding and Morrison 2013) or SCF (Ding et al. 2012; Comazzetto et al. 2019) was deleted from Tie2$^+$ endothelial cells using Tie2-Cre; CXCL12$^{flox/flox}$ or Tie2-Cre; SCF$^{flox/flox}$ mice, respectively. In contrast, other studies have shown that the number of GMPs was reduced but HSC number was not significantly reduced in CXCL12$^{flox/flox}$ mice crossed with Tie-2-Cre transgenic mice (Greenbaum et al. 2013; Agarwal et al. 2019).

Of note, endothelial cells are essential for emergency granulopoiesis induced by administration of lipopolysaccharide (LPS) modeling severe bacterial infection (Boettcher et al. 2014). Toll-like receptor (TLR) signal transduction occurs via adaptor molecules, including myeloid differentiation primary response gene 88 (Myd88) after LPS treatment. Although the number of granulocytes was markedly reduced in the bone marrow and.

increased in peripheral blood in LPS-injected wild-type mice, Tie2-Cre; Myd88$^{flox/flox}$ mice lacked the response, indicating that LPS acts on endothelial cells to induce emergency granulopoiesis (Boettcher et al. 2014).

6 Myh11$^+$NG2$^+$ Periarteriolar Cells

It has been reported that Nestin-GFP^{high}NG2$^+$ cells are in contact with arteries in the bone marrow from transgenic mice expressing GFP under the control of the neural-specific regulatory elements of Nestin gene (Kunisaki et al. 2013; Mignone et al. 2004) (Fig. 4). They have shown that Nestin-GFP^{high}NG2$^+$ cells contain cells that form CFU-Fs (Kunisaki et al. 2013), that the mean distance of CD150$^+$CD48$^-$Lin$^-$ HSCs to arterioles (52 μm) was highly statistically different from that of randomly placed CD150$^+$CD48$^-$Lin$^-$ HSCs (78 μm). Additionally, HSC numbers were reduced in the marrow after ablation of NG2-expressing cells in vivo using mice expressing DTR under the control of the regulatory elements of NG2 gene (NG2-CreERT; iDTR mice). Zhou et al. revealed that NG2 was also expressed in osteoblasts lining bone surfaces and nonmyelinating schwann cells lining arteries (Zhou et al. 2014) and that HSC number was unaltered when CXCL12 or SCF was deleted from Nestin-expressing cells or NG2-expressing cells using Nestin-Cre (Ding et al. 2012; Ding and Morrison 2013) or NG2-CreERT (Acar et al. 2015) mice. Asada et al. address this and show that the number of phenotypic HSCs was reduced by about twofold when CXCL12 was deleted from Myh11$^+$ cells, which are associated with arterioles and included in Nestin-GFP$^+$ cells, using Myh11-CreERT2; CXCL12$^{flox/flox}$ mice, concluding that Myh11$^+$NG2$^+$ periarteriolar cells were involved in HSC maintenance in the bone marrow (Asada et al. 2017). However, immunohistochemical analysis have shown that the localization of α-catulin-GFP$^+$c-kit$^+$ HSCs and H2B-GFP$^+$c-kit$^+$ HSCs relative to arterioles did not significantly

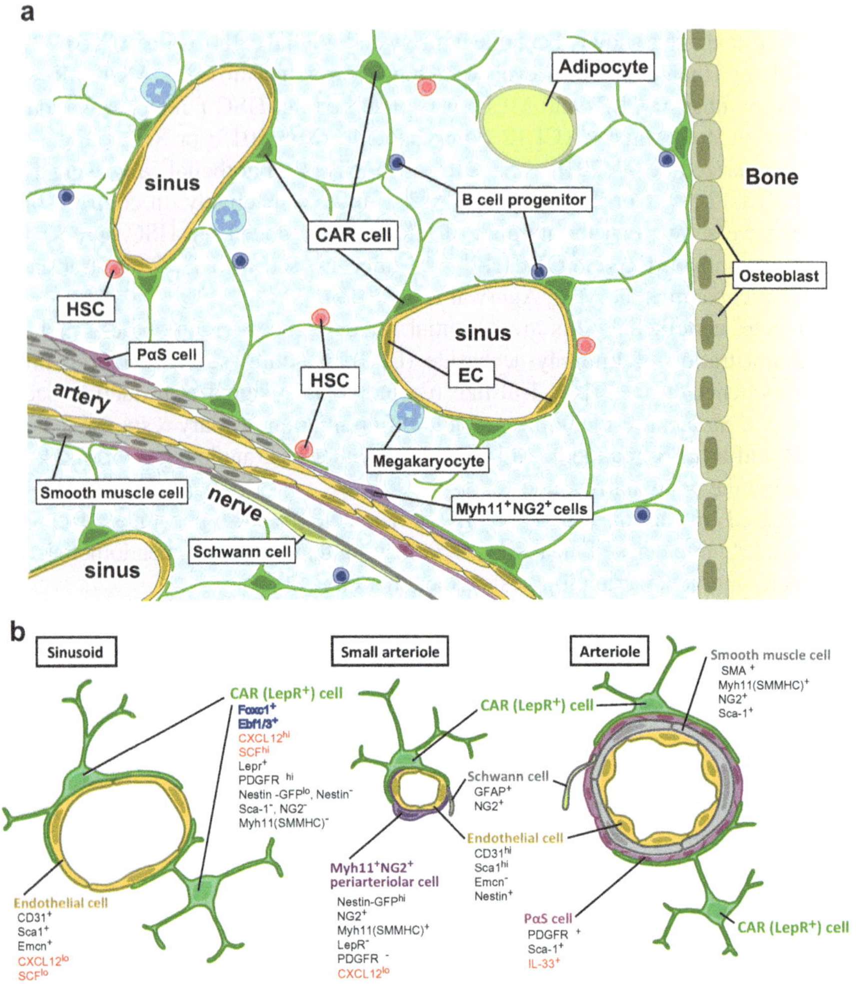

Fig. 4 Candidate cellular niches for HSCs and immune cells in bone marrow. HSCs are in contact with their cellular niches (**a**), including various types of perivascular cells, such as CAR/LepR$^+$ cells, endothelial cells, Myh11$^+$NG2$^+$ periarteriolar cells, PαS cells, and GFAP$^+$ nonmyelinating schwann cells (**a** and **b**).

differ from the distribution of random spots (Shimoto et al. 2017; Foudi et al. 2009). Furthermore, intravital imaging of HSCs using HSC-specific reporter mice with the GFP-Mds1-evi1 gene fusion product revealed that Mds1-evi1$^+$Flt3$^-$ HSCs were almost exclusively associated with sinusoids rather than arterioles (Christodoulou et al. 2020).

7 PDGFRα$^+$Sca-1$^+$CD45$^-$Ter119$^-$ (PαS) Cells

Morikawa et al. reported that PDGFRα$^+$Sca-1$^+$CD31$^-$CD45$^-$Ter119$^-$ (PαS) cells have potential to differentiate into adipocytes and osteoblasts, which are located along arteries in bone marrow (Morikawa et al. 2009) (Fig. 4). PαS cells are distinct from CAR cells because CAR cells express PDGFRα but not Sca-1 (Omatsu et al. 2010). Recently, Gomariz et al. have shown that PαS cells express higher levels of IL-33 than any other population of bone marrow (Helbling et al. 2019). When CXCL12 was conditionally deleted from PαS cells, all CAR cells, and osteoblasts, the numbers of LT-HSCs were severely reduced in bone marrow (Greenbaum et al. 2013). In contrast, when CXCL12 was conditionally deleted from the majority of CAR cells and osteoblasts, a modest decrease in LT-HSC number was observed. Based on these findings, Greenbaum et al. reported that CXCL12 produced by PαS cells was involved in HSC maintenance (Greenbaum et al. 2013). However, these findings do not rule out the possibility that CXCL12 production from a small subset of CAR cells contributes to HSC maintenance. Studies using mice, in which CXCL12 was conditionally deleted from PαS cells, will provide direct evidence for a role of PαS cells in HSC maintenance.

8 GFAP$^+$ Nonmyelinating Schwann Cells

Yamazaki et al. have reported that integrin-β8 expressed in glial fibrillary acidic protein (GFAP)$^+$nestin$^+$PDGFRα$^-$ nonmyelinating Schwann cells, which ensheath neural axons, are located near arteries and activate latent TGF-β for the maintenance of HSCs in bone marrow (Yamazaki et al. 2011). However, deep confocal imaging of α-catulin-GFP$^+$c-kit$^+$ HSCs show that HSCs rarely contact with GFAP$^+$ Schwann cells or nerve fibers and did not significantly differ from random spots in their distance from GFAP$^+$ Schwann cells (Acar et al. 2015). Conditional knockout of integrin-β8 in GFAP-expressing cells will provide direct evidence for a role of GFAP$^+$ nonmyelinating Schwann cells in HSC maintenance.

9 Megakaryocytes and αSMA$^+$ Macrophages

It has been reported that some HSCs are associated with megakaryocytes (Bruns et al. 2014; Pinho et al. 2018; Zhao et al. 2014; Nakamura-Ishizu et al. 2015), which are always found in close contact with bone marrow sinusoids (Junt et al. 2007), that thrombopoietin (TPO) is produced by megakaryocytes (Nakamura-Ishizu et al. 2015), and that ablation of megakaryocytes or conditional deletion of TGFβ1 in megakaryocytes led to activation of quiescent HSCs, suggesting the role of megakaryocytes in maintenance of HSC quiescence (Bruns et al. 2014; Pinho et al. 2018).

However, reduction in the number of platelets might enhance production of TPO in liver and/or kidney, which travels away and enters bone marrow cavities through diffusion, increasing HSC number expression in liver, suggesting an indirect effect of megakaryocytes on HSC maintenance. Consistent with this, Decker et al. have demonstrated that deletion of TPO from hepatocytes lead to marked reduction of HSCs (Decker et al. 2018), indicating that liver-derived TPO in addition to bone marrow-derived CXCL12 and SCF is required for HSC maintenance. In addition, immunohistochemical analysis has shown that H2B-LR-HSCs are not located in proximity to megakaryocytes (Saçma et al. 2019). On the other hand, it has been shown that some HSCs are associated with a subset of αSMA^+ monocytes and macrophages (Ludin et al. 2012); however, there is no compelling evidence that αSMA^+ monocytes and macrophages act as a niche for HSCs. Together, further studies will be needed to determine the role of megakaryocytes or αSMA^+ monocytes and macrophages in HSC regulation.

10 Conclusion

Among various types of cells in the bone marrow, the bulk of evidence has suggested that the population of mesenchymal stem cells, termed CAR cells, which overlap strongly with $LepR^+$ cells, is the major cellular component of niches for HSC maintenance and lympho-hematopoiesis (Sugiyama et al. 2006; Omatsu et al. 2010, 2014; Ding et al. 2012; Shimoto et al. 2017; Acar et al. 2015; Comazzetto et al. 2019; Cordeiro Gomes et al. 2016; Seike et al. 2018; Aoki et al. 2021; Gomariz et al. 2018) (Fig. 4). For example, (Zhang et al. 2003) most HSCs are in contact with $CAR/LepR^+$ cells. (Calvi et al. 2003) Selective ablation of $CAR/LepR^+$ cells in vivo led to a severe reduction in hematopoietic stem and progenitor cells (HSPCs) and immune cells. (Arai et al. 2004) Selective deletion of SCF from $CAR/LepR^+$ cells resulted in severe HSC reduction. (Kiel et al. 2007a) Selective deletion of the tanscription factor(s), Foxc1 or Ebf1/3 from $CAR/LepR^+$ cells resulted in severe reduction in HSPCs and immune cells.

It has been shown that cells that have potential to differentiate into adipocytes and osteoblasts like $CAR/LepR^+$ cells reside in virtually all post-natal organs and tissues (Silva et al. 2006). What are differences between these extramedullary fibroblastic cells and $CAR/LepR^+$ cells? Only CAR cells express much higher levels of multiple transcription factors essential for formation and maintenance of HSC niches, and cytokines critical for lympho-hematopoiesis than any other cells ($Foxc1^{high}Ebf3^{high}CXCL12^{high}SCF^{high}$ cells) (Fig. 1a), and provide the required number of osteoblasts lining the endosteum throughout the life span. In addition, $Foxl1^+$ telocytes, in which the transcription factor Foxl1 is specifically expressed, are the important source of Wnt proteins essential for homeostasis of the intestinal epithelium (Shoshkes-Carmel et al. 2018). These results suggest that the tissue-specific populations of fibroblastic reticular cells create niches for tissue stem cells and tissue homeostasis. How CAR/LepR$^+$ cells are generated during ontogeny is an important question for the future.

Several subclusters were identified within the CAR/LepR$^+$ cell population (Tikhonova et al. 2019; Baryawno et al. 2019; Wolock et al. 2019; Baccin et al. 2020). Of note, expression levels of many genes in CAR/LepR$^+$ cells might be upregurated or downregulated by cues from adjacent cells, including endothelial cells and hemtopoietic cells and some CAR/LepR$^+$ cells are differentiating into adipocytes or osteoblasts with distinct gene signatures. In addition to CAR/LepR$^+$ cells, various types of cells, including sinusoidal endothelial cells and periarteriolar Myh11$^+$NG2$^+$ cells, have been shown to be involved in creating niches for HSCs (Greenbaum et al. 2013; Ding and Morrison 2013; Asada et al. 2017; Méndez-Ferrer et al. 2010; Kunisaki et al. 2013). It will be interesting and important to determine more precisely the functions of these cells, including subclusters of CAR/LepR$^+$ cells in regulation of HSCs and lympho-hematopoiesis.

Together, recent findings revealing the nature and functions of various types of candidate cellular niches for HSCs provide basis for future mechanistic studies exploring the crosstalk between hematopoietic/immune cells and microenvironments in both health and disease.

References

Aoki K, Kurashige M, Ichii M, Higaki K, Sugiyama T, Kaito T, Ando W, Sugano N, Sakai T, Shibayama H, Takaori-Kondo A, Morii E, Kanakura Y, Nagasawa T (2021) Identification of CXCL12-abundant reticular cells in human adult bone marrow. Br J Haematol

Acar M, Kocherlakota KS, Murphy MM, Peyer JG, Oguro H, Inra CN, Jaiyeola C, Zhao Z, Luby-Phelps K, Morrison SJ (2015) Deep imaging of bone marrow shows non-dividing stem cells are mainly perisinusoidal. Nature 526:126–130

Agarwal P, Isringhausen S, Li H, Paterson AJ, He J, Gomariz Á, Nagasawa T, Nombela-Arrieta C, Bhatia R (2019) Mesenchymal niche-specific expression of Cxcl12 controls quiescence of treatment-resistant leukemia stem cells. Cell Stem Cell 24:769–784

Ara T, Tokoyoda K, Sugiyama T, Egawa T, Kawabata K, Nagasawa T (2003) Long-term hematopoietic stem cells require stromal cell-derived factor-1 for colonizing bone marrow during ontogeny. Immunity 19:257–267

Arai F, Hirao A, Ohmura M, Sato H, Matsuoka S, Takubo K, Ito K, Koh GY, Suda T (2004) Tie2/angiopoietin-1 signaling regulates hematopoietic stem cell quiescence in the bone marrow niche. Cell 118:149–161

Asada N, Kunisaki Y, Pierce H, Wang Z, Fernandez NF, Birbrair A, Ma'ayan A, Frenette PS (2017) Differential cytokine contributions of perivascular haematopoietic stem cell niches. Nat Cell Biol 19:214–223

Baccin C, Al-Sabah J, Velten L, Helbling PM, Grünschläger F, Hernández-Malmierca P, Nombela-Arrieta C, Steinmetz LM, Trumpp A, Haas S (2020) Combined single-cell and spatial transcriptomics reveal the molecular, cellular and spatial bone marrow niche organization. Nat Cell Biol 22:38–48

Barker JE (1994) SI/SI(d) hematopoietic progenitors are deficient in situ. Exp Hematol 22:174–177

Barker N, Van Es JH, Kuipers J, Kujala P, Van Den Born M, Cozijnsen M, Haegebarth A, Korving J, Begthel H, Peters PJ, Clevers H (2007) Identification of stem cells in small intestine and colon by marker gene Lgr5. Nature 449:1003–1007

Baryawno N, Przybylski D, Kowalczyk MS, Kfoury Y, Severe N, Gustafsson K, Kokkaliaris KD, Mercier F, Tabaka M, Hofree M, Dionne D, Papazian A, Lee D, Ashenberg O, Subramanian A, Vaishnav ED, Rozenblatt-Rosen O, Regev A, Scadden DT (2019) A Cellular Taxonomy of the Bone Marrow Stroma in Homeostasis and Leukemia. Cell 177:1915–1932

Boettcher S, Gerosa RC, Radpour R, Bauer J, Ampenberger F, Heikenwalder M, Kopf M, Manz MG (2014) Endothelial cells translate pathogen signals into G-CSF-driven emergency granulopoiesis. Blood 124:1393–1403

Bowers M, Zhang B, Ho Y, Agarwal P, Chen CC, Bhatia R (2015) Osteoblast ablation reduces normal long-term hematopoietic stem cell self-renewal but accelerates leukemia development. Blood 125:2678–2688

Bruns I, Lucas D, Pinho S, Ahmed J, Lambert MP, Kunisaki Y, Scheiermann C, Schiff L, Poncz M, Bergman A, Frenette PS (2014) Megakaryocytes regulate hematopoietic stem cell quiescence through CXCL4 secretion. Nat Med 20:1315–1320

Calvi LM, Adams GB, Weibrecht KW, Weber JM, Olson DP, Knight MC, Martin RP, Schipani E, Divieti P, Bringhurst FR, Milner LA, Kronenberg HM, Scadden DT (2003) Osteoblastic cells regulate the haematopoietic stem cell niche. Nature 425:841–846

Cattoretti G, Schiro R, Orazi A, Soligo D, Colombo MP (1993) Bone marrow stroma in humans: anti-nerve growth factor receptor antibodies selectively stain reticular cells in vivo and in vitro. Blood 81:1726–1738

Christodoulou C, Spencer JA, Yeh SCA, Turcotte R, Kokkaliaris KD, Panero R, Ramos A, Guo G, Seyedhassantehrani N, Esipova TV, Vinogradov SA, Rudzinskas S, Zhang Y, Perkins AS, Orkin SH, Calogero RA, Schroeder T, Lin CP, Camargo FD (2020) Live-animal imaging of native haematopoietic stem and progenitor cells. Nature 578:278–283

Churchman SM, Ponchel F, Boxall SA, Cuthbert R, Kouroupis D, Roshdy T, Giannoudis PV, Emery P, McGonagle D, Jones EA (2012) Transcriptional profile of native CD271+ multipotential stromal cells: evidence for multiple fates, with prominent osteogenic and wnt pathway signaling activity. Arthritis Rheum 64:2632–2643

Comazzetto S, Murphy MM, Berto S, Jeffery E, Zhao Z, Morrison SJ (2019) Restricted hematopoietic progenitors and erythropoiesis require SCF from leptin receptor+ niche cells in the bone marrow. Cell Stem Cell 24:477–486

Cordeiro Gomes A, Hara T, Lim VY, Herndler-Brandstetter D, Nevius E, Sugiyama T, Tani-ichi S, Schlenner S, Richie E, Rodewald HR, Flavell RA, Nagasawa T, Ikuta K, Pereira JP (2016) Hematopoietic stem cell niches produce lineage-instructive signals to control multipotent progenitor differentiation. Immunity 45:1219–1231

Crozatier M, Ubeda JM, Vincent A, Meister M (2004) Cellular immune response to parasitization in Drosophila requires the EBF orthologue collier. PLoS Biol 2:E196

da Silva ML, Chagastelles PC, Nardi NB (2006) Mesenchymal stem cells reside in virtually all post-natal organs and tissues. J Cell Sci 119:2204–2213

De Bruyn PPH, Breen PC, Thomas TB (1970) The microcirculation of the bone marrow. Anat Rec 168:55–68

Decker M, Leslie J, Liu Q, Ding L (2018) Hepatic thrombopoietin is required for bone marrow hematopoietic stem cell maintenance. Science 360:106–110

Decker M, Martinez-Morentin L, Wang G, Lee Y, Liu Q, Leslie J, Ding L (2017) Leptin-receptor-expressing bone marrow stromal cells are myofibroblasts in primary myelofibrosis. Nat Cell Biol 19:677–688

Degirmenci B, Valenta T, Dimitrieva S, Hausmann G, Basler K (2018) GLI1-expressing mesenchymal cells form the essential Wnt-secreting niche for colon stem cells. Nature 558:449–453

Dexter TM, Allen TD, Lajtha LG (1977) Conditions controlling the proliferation of haemopoietic stem cells in vitro. J Cell Physiol 91:335–344

Ding L, Morrison SJ (2013) Haematopoietic stem cells and early lymphoid progenitors occupy distinct bone marrow niches. Nature 495:231–235

Ding L, Saunders TL, Enikolopov G, Morrison SJ (2012) Endothelial and perivascular cells maintain haematopoietic stem cells. Nature 481:457–462
Durand A, Donahue B, Peignon G, Letourneur F, Cagnard N, Slomianny C, Perret C, Shroyer NF, Romagnolo B (2012) 'Functional intestinal stem cells after Paneth cell ablation induced by the loss of transcription factor Math1 (Atoh1). Proc Natl Acad Sci U S A 109:8965–8970
Flores-Figueroa E, Varma S, Montgomery K, Greenberg PL, Gratzinger D (2012) Distinctive contact between CD34+ hematopoietic progenitors and CXCL12+ CD271+ mesenchymal stromal cells in benign and myelodysplastic bone marrow. Lab Investig 92:1330–1341
Foudi A, Hochedlinger K, Van Buren D, Schindler JW, Jaenisch R, Carey V, Hock H (2009) Analysis of histone 2B-GFP retention reveals slowly cycling hematopoietic stem cells. Nat Biotechnol 27:84–90
Friedenstein AJ, Chailakhyan RK, Latsinik N V., Panasyvk AF, Keiliss-Borok IV (1974) Stromal cells responsible for transferring the microenvironment of the hemopoietic tissues: Cloning in vitro and retransplantation in vivo. Transplantation 17:331–340
Gomariz A, Helbling PM, Isringhausen S, Suessbier U, Becker A, Boss A, Nagasawa T, Paul G, Goksel O, Székely G, Stoma S, Nørrelykke SF, Manz MG, Nombela-Arrieta C (2018) Quantitative spatial analysis of haematopoiesis-regulating stromal cells in the bone marrow microenvironment by 3D microscopy. Nat Commun 9:2532
Greenbaum A, Hsu YMS, Day RB, Schuettpelz LG, Christopher MJ, Borgerding JN, Nagasawa T, Link DC (2013) CXCL12 in early mesenchymal progenitors is required for haematopoietic stem-cell maintenance. Nature 495:227–230
Helbling PM, Piñeiro-Yáñez E, Gerosa R, Boettcher S, Al-Shahrour F, Manz MG, Nombela-Arrieta C (2019) Global transcriptomic profiling of the bone marrow stromal microenvironment during postnatal development, aging, and inflammation. Cell Rep 29:3313–3330
Hooper AT, Butler JM, Nolan DJ, Kranz A, Iida K, Kobayashi M, Kopp HG, Shido K, Petit I, Yanger K, James D, Witte L, Zhu Z, Wu Y, Pytowski B, Rosenwaks Z, Mittal V, Sato TN, Rafii S (2009) Engraftment and reconstitution of hematopoiesis is dependent on VEGFR2-mediated regeneration of sinusoidal endothelial cells. Cell Stem Cell 4:263–274
Jones EA, Kinsey SE, English A, Jones RA, Straszynski L, Meredith DM, Markham AF, Jack A, Emery P, McGonagle D (2002) Isolation and characterization of bone marrow multipotential mesenchymal progenitor cells. Arthritis Rheum 46:3349–3360
Junt T, Schulze H, Chen Z, Massberg S, Goerge T, Krueger A, Wagner DD, Graf T, Italiano JE, Shivdasani RA, Von Andrian UH (2007) Dynamic visualization of thrombopoiesis within bone marrow. Science 317:1767–1770
Kabiri Z, Greicius G, Madan B, Biechele S, Zhong Z, Zaribafzadeh H, Edison AJ, Wu Y, Bunte R, Williams BO, Rossant J, Virshup DM (2014) Stroma provides an intestinal stem cell niche in the absence of epithelial Wnts. Development 141:2206–2215
Karsunky H, Inlay MA, Serwold T, Bhattacharya D, Weissman IL (2008) Flk2 + common lymphoid progenitors possess equivalent differentiation potential for the B and T lineages. Blood 111:5562–5570
Kiel MJ, Yilmaz ÖH, Iwashita T, Yilmaz OH, Terhorst C, Morrison SJ (2005) SLAM family receptors distinguish hematopoietic stem and progenitor cells and reveal endothelial niches for stem cells. Cell 121:1109–1121
Kiel MJ, Radice GL, Morrison SJ (2007a) Lack of evidence that hematopoietic stem cells depend on N-cadherin-mediated adhesion to osteoblasts for their maintenance. Cell Stem Cell 1:204–217
Kiel MJ, He S, Ashkenazi R, Gentry SN, Teta M, Kushner JA, Jackson TL, Morrison SJ (2007b) Haematopoietic stem cells do not asymmetrically segregate chromosomes or retain BrdU. Nature 449:238–242
Kim TH, Escudero S, Shivdasani RA (2012) Intact function of Lgr5 receptor-expressing intestinal stem cells in the absence of Paneth cells. Proc Natl Acad Sci U S A 109:3932–3937
Kohara H, Omatsu Y, Sugiyama T, Noda M, Fujii N, Nagasawa T (2007) Development of plasmacytoid dendritic cells in bone marrow stromal cell niches requires CXCL12-CXCR4 chemokine signaling. Blood 110:4153–4160

Korinek V, Barker N, Moerer P, Van Donselaar E, Huls G, Peters PJ, Clevers H (1998) Depletion of epithelial stem-cell compartments in the small intestine of mice lacking Tcf-4. Nat Genet 19:379–383

Kume T, Deng KY, Winfrey V, Gould DB, Walter MA, Hogan BLM (1998) The forkhead/winged helix gene Mf1 is disrupted in the pleiotropic mouse mutation congenital hydrocephalus. Cell 93:985–996

Kunisaki Y, Bruns I, Scheiermann C, Ahmed J, Pinho S, Zhang D, Mizoguchi T, Wei Q, Lucas D, Ito K, Mar JC, Bergman A, Frenette PS (2013) Arteriolar niches maintain haematopoietic stem cell quiescence. Nature 502:637–643

Li H, Ghazanfari R, Zacharaki D, Ditzel N, Isern J, Ekblom M, Méndez-Ferrer S, Kassem M, Scheding S (2014) Low/negative expression of PDGFR-α identifies the candidate primary mesenchymal stromal cells in adult human bone marrow. Stem Cell Rep 3:965–974

Ludin A, Itkin T, Gur-Cohen S, Mildner A, Shezen E, Golan K, Kollet O, Kalinkovich A, Porat Z, D'Uva G, Schajnovitz A, Voronov E, Brenner DA, Apte RN, Jung S, Lapidot T (2012) Monocytes-macrophages that express α-smooth muscle actin preserve primitive hematopoietic cells in the bone marrow. Nat Immunol 13:1072–1082

Méndez-Ferrer S, Michurina T V., Ferraro F, Mazloom AR, MacArthur BD, Lira SA, Scadden DT, Ma'ayan A, Enikolopov GN, Frenette PS (2010) Mesenchymal and haematopoietic stem cells form a unique bone marrow niche. Nature 466:829–834

Mignone JL, Kukekov V, Chiang AS, Steindler D, Enikolopov G (2004) Neural stem and progenitor cells in Nestin-GFP transgenic mice. J Comp Neurol 469:311–324

Miller CL, Rebel VI, Lemieux ME, Helgason CD, Lansdorp PM, Eaves CJ (1996) Studies of W mutant mice provide evidence for alternate mechanisms capable of activating hematopoietic stem cells. Exp Hematol 24:185–194

Morikawa S, Mabuchi Y, Kubota Y, Nagai Y, Niibe K, Hiratsu E, Suzuki S, Miyauchi-Hara C, Nagoshi N, Sunabori T, Shimmura S, Miyawaki A, Nakagawa T, Suda T, Okano H, Matsuzaki Y (2009) Prospective identification, isolation, and systemic transplantation of multipotent mesenchymal stem cells in murine bone marrow. J Exp Med 206:2483–2496

Nagasawa T, Hirota S, Tachibana K, Takakura N, Nishikawa SI, Kitamura Y, Yoshida N, Kikutani H, Kishimoto T (1996) Defects of B-cell lymphopoiesis and bone-marrow myelopoiesis in mice lacking the CXC chemokine PBSF/SDF-1. Nature 382:635–638

Nakahara F, Borger DK, Wei Q, Pinho S, Maryanovich M, Zahalka AH, Suzuki M, Cruz CD, Wang Z, Xu C, Boulais PE, Ma'ayan A, Greally JM, Frenette PS (2019) Engineering a haematopoietic stem cell niche by revitalizing mesenchymal stromal cells. Nat Cell Biol 21:560–567

Nakamura-Ishizu A, Takubo K, Kobayashi H, Suzuki-Inoue K, Suda T (2015) CLEC-2 in megakaryocytes is critical for maintenance of hematopoietic stem cells in the bone marrow. J Exp Med 212:2133–2146

Naveiras O, Nardi V, Wenzel PL, Hauschka PV, Fahey F, Daley GQ (2009) Bone-marrow adipocytes as negative regulators of the haematopoietic microenvironment. Nature 460:259–263

Noda M, Omatsu Y, Sugiyama T, Oishi S, Fujii N, Nagasawa T (2011) CXCL12-CXCR4 chemokine signaling is essential for NK-cell development in adult mice. Blood 117:451–458

Omatsu Y, Sugiyama T, Kohara H, Kondoh G, Fujii N, Kohno K, Nagasawa T (2010) The essential functions of adipo-osteogenic progenitors as the hematopoietic stem and progenitor cell niche. Immunity 33:387–399

Omatsu Y, Seike M, Sugiyama T, Kume T, Nagasawa T (2014) Foxc1 is a critical regulator of haematopoietic stem/progenitor cell niche formation. Nature 508:536–540

Pinho S, Marchand T, Yang E, Wei Q, Nerlov C, Frenette PS (2018) Lineage-biased hematopoietic stem cells are regulated by distinct niches. Dev Cell 44:634–641

Pittenger MF, Mackay AM, Beck SC, Jaiswal RK, Douglas R, Mosca JD, Moorman MA, Simonetti DW, Craig S, Marshak DR (1999) Multilineage potential of adult human mesenchymal stem cells. Science 284:143–147

Quirici N, Soligo D, Bossolasco P, Servida F, Lumini C, Deliliers GL (2002) Isolation of bone marrow mesenchymal stem cells by anti-nerve growth factor receptor antibodies. Exp Hematol 30:783–791

Sacchetti B, Funari A, Michienzi S, Di Cesare S, Piersanti S, Saggio I, Tagliafico E, Ferrari S, Robey PG, Riminucci M, Bianco P (2007) Self-renewing osteoprogenitors in bone marrow sinusoids can organize a hematopoietic microenvironment. Cell 131:324–336

Saçma M, Pospiech J, Bogeska R, de Back W, Mallm JP, Sakk V, Soller K, Marka G, Vollmer A, Karns R, Cabezas-Wallscheid N, Trumpp A, Méndez-Ferrer S, Milsom MD, Mulaw MA, Geiger H, Florian MC (2019) Haematopoietic stem cells in perisinusoidal niches are protected from ageing. Nat Cell Biol 21:1309–1320

Sangiorgi E, Capecchi MR (2008) Bmi1 is expressed in vivo in intestinal stem cells. Nat Genet 40:915–920

Sato T, Van Es JH, Snippert HJ, Stange DE, Vries RG, Van Den Born M, Barker N, Shroyer NF, Van De Wetering M, Clevers H (2011) Paneth cells constitute the niche for Lgr5 stem cells in intestinal crypts. Nature 469:415–418

Seike M, Omatsu Y, Watanabe H, Kondoh G, Nagasawa T (2018) Stem cell niche-specific Ebf3 maintains the bone marrow cavity. Genes Dev 32:359–372

Shimoto M, Sugiyama T, Nagasawa T (2017) Numerous niches for hematopoietic stem cells remain empty during homeostasis. Blood 129:2124–2131

Shoshkes-Carmel M, Wang YJ, Wangensteen KJ, Tóth B, Kondo A, Massassa EE, Itzkovitz S, Kaestner KH (2018) Subepithelial telocytes are an important source of Wnts that supports intestinal crypts. Nature 557:242–246

Sugiyama T, Kohara H, Noda M, Nagasawa T (2006) Maintenance of the hematopoietic stem cell pool by CXCL12-CXCR4 chemokine signaling in bone marrow stromal cell niches. Immunity 25:977–988

Tachibana K, Hirota S, Iizasa H, Yoshida H, Kawabata K, Kataoka Y, Kitamura Y, Matsushima K, Yoshida N, Nishikawa SI, Kishimoto T, Nagasawa T (1998) The chemokine receptor CXCR4 is essential for vascularization of the gastrointestinal tract. Nature 393:591–594

Terashima A, Okamoto K, Nakashima T, Akira S, Ikuta K, Takayanagi H (2016) Sepsis-induced osteoblast ablation causes immunodeficiency. Immunity 44:1434–1443

Tikhonova AN, Dolgalev I, Hu H, Sivaraj KK, Hoxha E, Cuesta-Domínguez Á, Pinho S, Akhmetzyanova I, Gao J, Witkowski M, Guillamot M, Gutkin MC, Zhang Y, Marier C, Diefenbach C, Kousteni S, Heguy A, Zhong H, Fooksman DR, Butler JM, Economides A, Frenette PS, Adams RH, Satija R, Tsirigos A, Aifantis I (2019) The bone marrow microenvironment at single-cell resolution. Nature 569:222–228

Tormin A, Li O, Brune JC, Walsh S, Schütz B, Ehinger M, Ditzel N, Kassem M, Scheding S (2011) CD146 expression on primary nonhematopoietic bone marrow stem cells is correlated with in situ localization. Blood 117:5067–5077

Upadhaya S, Krichevsky O, Akhmetzyanova I, Sawai CM, Fooksman DR, Reizis B (2020) Intravital imaging reveals motility of adult hematopoietic stem cells in the bone marrow niche. Cell Stem Cell 27:336–345

Wolock SL, Krishnan I, Tenen DE, Matkins V, Camacho V, Patel S, Agarwal P, Bhatia R, Tenen DG, Klein AM, Welner RS (2019) Mapping distinct bone marrow niche populations and their differentiation paths. Cell Rep 28:302–311

Yamazaki S, Ema H, Karlsson G, Yamaguchi T, Miyoshi H, Shioda S, Taketo MM, Karlsson S, Iwama A, Nakauchi H (2011) Nonmyelinating schwann cells maintain hematopoietic stem cell hibernation in the bone marrow niche. Cell 147:1146–1158

Yu VWC, Saez B, Cook C, Lotinun S, Pardo-Saganta A, Wang YH, Lymperi S, Ferraro F, Raaijmakers MHGP, Wu JY, Zhou L, Rajagopal J, Kronenberg HM, Baron R, Scadden DT (2015) Specific bone cells produce DLL4 to generate thymus-seeding progenitors from bone marrow. J Exp Med 212:759–774

Zhang B, Ho YW, Huang Q, Maeda T, Lin A, Lee SUK, Hair A, Holyoake TL, Huettner C, Bhatia R (2012) Altered microenvironmental regulation of leukemic and normal stem cells in chronic myelogenous leukemia. Cancer Cell 21:577–592

Zhang J, Niu C, Ye L, Huang H, He X, Tong WG, Ross J, Haug J, Johnson T, Feng JQ, Harris S, Wiedemann LM, Mishina Y, Li L (2003) Identification of the haematopoietic stem cell niche and control of the niche size. Nature 425:836–841

Zhang J, Link DC (2016) Targeting of mesenchymal stromal cells by cre-recombinase transgenes commonly used to target osteoblast lineage cells. J Bone Miner Res 31:2001–2007

Zhao M, Perry JM, Marshall H, Venkatraman A, Qian P, He XC, Ahamed J, Li L (2014) Megakaryocytes maintain homeostatic quiescence and promote post-injury regeneration of hematopoietic stem cells. Nat Med 20:1321–1326

Zhou BO, Yue R, Murphy MM, Peyer JG, Morrison SJ (2014) Leptin-receptor-expressing mesenchymal stromal cells represent the main source of bone formed by adult bone marrow. Cell Stem Cell 15:154–168

Zhou BO, Ding L, Morrison SJ (2015) Hematopoietic stem and progenitor cells regulate the regeneration of their niche by secreting Angiopoietin-1. Elife 4:e05521

Zhou BO, Yu H, Yue R, Zhao Z, Rios JJ, Naveiras O, Morrison SJ (2017) Bone marrow adipocytes promote the regeneration of stem cells and haematopoiesis by secreting SCF. Nat Cell Biol 19:891–903

Zou YR, Kottman AH, Kuroda M, Taniuchi I, Littman DR (1998) Function of the chemokine receptor CXCR4 in heaematopolesis and in cerebellar development. Nature 393:595–599

Remodeling of the Bone Marrow Stromal Microenvironment During Pathogenic Infections

YeVin Mun, Serena Fazio, and César Nombela Arrieta

Abstract The bone marrow (BM) is the primary hematopoietic organ and a hub in which organismal demands for blood cellular output are systematically monitored. BM tissues are additionally home to a plethora of mature immune cell types, providing functional environments for the activation of immune responses and acting as preferred anatomical reservoirs for cells involved in immunological memory. Stromal cells of the BM microenvironment crucially govern different aspects of organ function, by structuring tissue microanatomy and by directly providing essential regulatory cues to hematopoietic and immune components in distinct niches. Emerging evidence demonstrates that stromal networks are endowed with remarkable functional and structural plasticity. Stress-induced adaptations of stromal cells translate into demand-driven hematopoiesis. Furthermore, aberrations of stromal integrity arising from pathological conditions critically contribute to the dysregulation of BM function. Here, we summarize our current understanding of the alterations that pathogenic infections and ensuing inflammatory conditions elicit on the global topography of the BM microenvironment, the integrity of anatomical niches and cellular interactions, and ultimately, on the regulatory function of diverse stromal subsets.

1 Functional Specialization of BM In Anatomical Niches

1.1 Structure And Function In BM Tissues

The bone marrow (BM) is home to a myriad of fast-paced cellular processes, which result in billions of mature hematopoietic cells being released into the peripheral circulation every minute (Nombela-Arrieta and Manz 2017). Beyond facilitating the continuous differentiation of primitive progenitor cells, the tissue environment of the

---

Y. Mun · S. Fazio · C. N. Arrieta (✉)
Department of Medical Oncology and Hematology, University and University Hospital Zurich, Häldeliweg 4, 8032 Zurich, Switzerland
e-mail: Cesar.NombelaArrieta@usz.ch

T. Nagasawa (ed.), *Bone Marrow Niche*, Current Topics in Microbiology and Immunology 434, https://doi.org/10.1007/978-3-030-86016-5_3

BM also nurtures the maintenance of an undifferentiated reservoir of hematopoietic stem cells (HSCs), whose regenerative capacity is activated and mobilized upon demand (Morrison and Scadden 2014; Takizawa et al. 2012). Less well studied but also crucial, is the ability of the BM to support the generation of cellular immune responses and the capacity to host the largest fractions of rare, yet essential immune cell populations, such as memory T cells (Tm), plasma cells (PCs) and regulatory T cells (Tregs) (Mercier et al. 2012; Feuerer et al. 2003; Tokoyoda et al. 2010). All these crucial, organ-specific functions are articulated around a relatively stable cellular infrastructure built by so-called stromal, non-hematopoietic components of endothelial, mesenchymal and neural origin, which are integral to the regulation of hematopoiesis and immune functions of the BM (Pinho and Frenette 2019; Kfoury and Scadden 2015). Perhaps most importantly, BM function is highly versatile and under constant adaptation to the changing demands for hematopoietic supply. Among the conditions that exert the highest functional pressure to BM tissues are infectious challenges, which cause drastic shifts in the dynamics of blood cell consumption and bring about potent compensatory responses in the BM (Zaretsky et al. 2013).

Despite the unparalleled knowledge of the hematopoietic system gained in the past years, how the cellular complexity and functional diversification of BM tissues are spatially organized and integrated in a defined anatomical context remains poorly understood. Secondary lymphoid organs, such as spleen, peripheral lymph nodes (PLN) and Peyer's patches (PP) are characterized by their canonical and well-described anatomy. Specific cell types occupy designated regions, typically demarcated by the presence of different stromal cells, which have relatively fixed locations within the global tissue structure (Junt et al. 2008). In contrast, defined compartmentalization of cells or developmental processes has not been clearly observed in the BM. Current models suggest that cell fate decisions rely mostly on interactions established between hematopoietic and stromal cells within microniches, which can be defined by two criteria: (i) the specific and non-stochastic spatial association of a limited number (two or more) of cell subsets and (ii) the establishment between these individual components of mutualistic relationships based on an intense molecular crosstalk, which are of functional relevance to hematopoiesis or immune function (Mercier et al. 2012; Crane et al. 2017; Wei and Frenette 2018). Multiple examples of such domains have been visualized and their roles analyzed. Nonetheless, the global topography within the organ-wide microarchitecture, according to which these functional nodes may be assembled, and the potential hierarchical communication between them have not been widely explored. Of importance, akin to other organs, evidence now suggests that pathological states of BM function often correlate with the intense disruption of the tissue structure and the consequent perturbation of the physical associations, which directly impact the functionality of microniches (Mueller and Germain 2009; Méndez-Ferrer et al. 2020).

In this review, we aim to outline a schematic picture of our current knowledge on the organization of tissue landscape in the BM, and describe how inflammatory conditions driven by pathogenic infections modify this native homeostatic configuration. A unique property of the BM is that while working as a single organ, it is disseminated throughout physically distant locations. Studies suggest that BM in different osseous cavities fundamentally share a similar composition in steady state (Kiel et al. 2005; Lassailly et al. 2013). Hence, it is here assumed that the niches described so far, mostly in femoral or sternal murine BM tissues, are universally and homogeneously found across BM cavities. However, certain conditions such as aging are known to modify structure and composition differentially between BM sites. Therefore, it cannot be ruled out that a similar phenomenon may be observed in infectious processes, especially those that progress chronically and focally inside specific organs or tissues.

1.2 The Paradigm of The HSC Niche

Since the original postulation of the HSC niche in 1978 (Schofield 1978), the identification of the microanatomical domains in which HSCs reside and the characterization of the minimal cellular and molecular determinants that define their maintenance have historically attracted the highest interest and inspired studies on BM anatomy (Morrison and Scadden 2014). Initial evidence suggested that HSCs preferentially concentrated in bone proximal and endosteal regions, and were regulated by mature osteoblastic cells. However, the refinement of imaging technologies, as well as of the specificity of cellular labeling strategies, have allowed to unequivocally map HSCs mostly scattered throughout BM topography (Crane et al. 2017). Over recent years, successive studies have functionally associated HSCs with a number of BM cell types including (i) mesenchymal populations such as CXCL12-abundant reticular (CARc) and Nestin-GFP^{hi} cells (Omatsu et al. 2010; Kunisaki et al. 2013; Acar et al. 2015), (ii) vascular networks of arterial and sinusoidal nature (Kunisaki et al. 2013; Acar et al. 2015; Itkin et al. 2016) and (iii) different hematopoietic subsets, such as megakaryocytes (Bruns et al. 2014; Zhao 2014). Genetic targeting or straightforward depletion of these cell types induces immediate and measurable effects in the numbers or cycling status of HSCs, which together with their observed spatial relationships in microscopy studies (Gomariz et al. 2019) have led to the consideration of these cell types as putative components of complex multicellular HSC niches. However, whether interventions in single individual niche cells affect HSCs in a direct manner, or do so indirectly by influencing other critical cellular compartments, or stages of the hematopoietic hierarchy that in turn activate/alter HSCs, is unclear for certain proposed niche components. Thus, the exact mechanisms by which certain of these cell types contribute to the control of HSC homeostasis remain elusive, and our perception of the HSC niche is under constant redefinition. For nuanced and comprehensive discussions on the current state of the field, we refer the reader to excellent published reviews on this topic (Pinho and Frenette 2019; Crane et al. 2017).

Most importantly, and despite the referred uncertainties, it is widely accepted that the vast majority of HSCs reside in the proximity of sinusoidal microvasculature, a dense highly interconnected labyrinth-like tree, which carries deoxygenated, venous blood and BM egressing cells back into the systemic circulation (Fig. 1) (Sivaraj

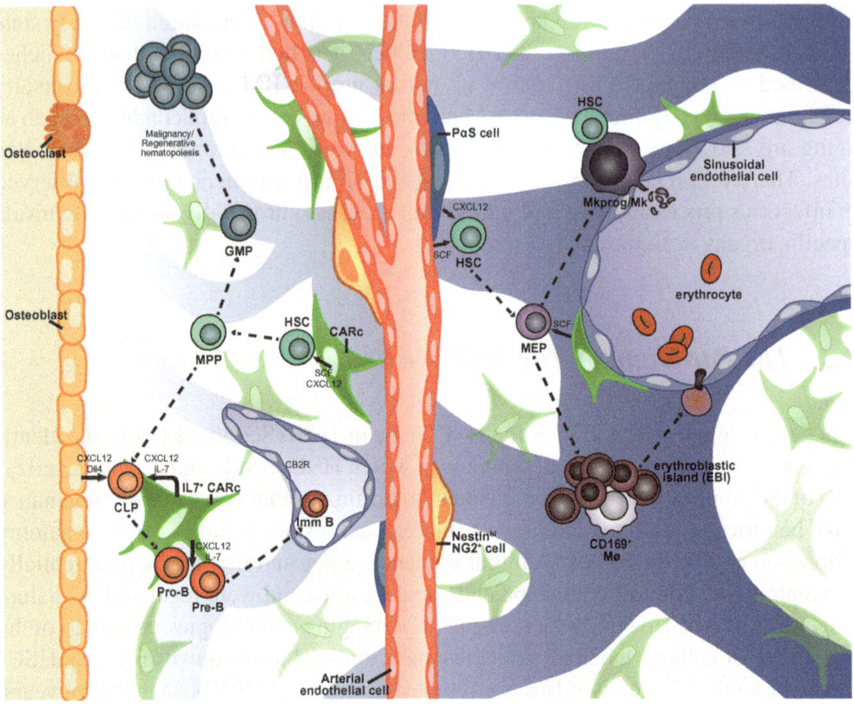

Fig. 1 **HSC and Progenitor cell and niches in the BM microenvironment**. A schematic view of the interactions established by HSPCs and stromal cell networks, as described in the text. Discontinuous arrows depict hierarchical relationships between cell types. Continuous arrows indicate anatomical or functional relationships. Factors involved in cell–cell communication are shown. **From left to right**: A fraction of CLPs interact with bone-lining osteoblasts via low production of CXCL12, while most CLPS rely on interactions with IL-7^{+}CARc. B cell developmental trajectories depend on the sustained provision of IL-7 by a subset of CARc, while terminal maturation stages in the BM take place within the intraluminal side of sinusoidal endothelial walls (bottom). GMPs are found scattered in the BM, depend on interactions with CARc and aggregate during stress-induced hematopoietic responses (top). Center: a simplified scheme of HSC niches; the vast majority of HSCs are found in contact with sinusoidal vasculature (blue) and CARc, while a small fraction resides in the vicinity of arterial/arteriolar vessels (red) in contact with Nestin^{hi} mesenchymal cells and SCF-producing AECs. HSCs have also been observed to interact with megakaryocytes. MEPs lie adjacent to CARc, and depend on CARc-derived SCF. Differentiated erythroid precursors are depicted within an erythroblastic island, in contact with a nurturing macrophage

and Adams 2016). Within these perivascular regions, HSCs lie adjacent to sinusoidal endothelial cells (SECs) and CARc, which form a heterogeneous subset of mesenchymal progenitors, with the potential to give rise to adipocytes and osteoblasts (Omatsu et al. 2010; Sugiyama et al. 2006). SECs and CARc are absolutely essential for HSC maintenance. On the one hand, SECs support HSCs through the production of a number of different angiocrine molecules, mainly Stem Cell Factor (SCF) (Ding et al. 2012). CARc, in turn, are the fundamental source of factors such as CXCL12, PTN (pleiothropin) and also SCF (Ding et al. 2012; Ding and Morrison

2013; Greenbaum et al. 2013; Himburg et al. 2018). Subset-specific deletion of any of these factors, or direct ablation of the entire CARc population, leads to significant changes in HSC numbers, changes in localization toward extramedullary sites and/or exit of homeostatic quiescence (Omatsu et al. 2010,2014; Ding et al. 2012; Ding and Morrison 2013; Greenbaum et al. 2013; Himburg et al. 2018). However, it is important to note that sinusoids profusely occupy BM spaces and CARc are extremely abundant, and ensheath virtually any segment of the sinusoidal network (Gomariz et al. 2018). Thus, rather than a restricted location, CARc-rich perisinusoidal domains represent a widespread location of BM tissues in which not only HSCs but also the majority of hematopoietic cells transiently or permanently reside, even in the absence of non-stochastic interaction forces. In fact, the cytokine profile of CARc is very broad as it includes the main pro-lymphoid (*Il7, Il15 and Ccl19)* and pro-myeloid factors (*Csf1, Il34 and Ccl2*), for which they have been termed professional cytokine-producing cells (Baccin et al. 2019). It is therefore not surprising that CARc or subsets thereof control many levels of hematopoietic development, and are absolutely essential for the maintenance of virtually all progenitor cells, as will be discussed below. Notably, recent single-cell analyses have delineated the existence of at least 3–8 distinct subsets, with partially distinct gene expression profiles and differentiation biases (Baccin et al. 2019; Wolock et al. 2019; Baryawno 2019; Matsushita et al. 2020; Tikhonova et al. 2019). Therefore, the possibility remains that different subpopulations may display tropism and affinity for specific progenitor stages, as well as specialized functional properties, in which case a CARc subset-specific HSC niche could exist.

Finally, an alternative niche, which also relies on HSC interactions with a distinct endothelial–mesenchymal interface, has been proposed to safeguard quiescent HSCs (Kunisaki et al. 2013; Itkin et al. 2016). Arteriolar vessels are lined by SCF and CXCL12-producing arterial EC (AEC) cells, and are ensheathed by several layers of pericytes, which include PDGFR-α^+Sca-1^+ (PαSc) and *Ng2* $^+$/Nestin-GFP^{hi} (Kunisaki et al. 2013; Helbling et al. 2019; Xu 2018). A minor fraction of HSCs and progenitor cells (~10%) has been consistently observed to lie adjacent to these microvessels (Kunisaki et al. 2013; Acar et al. 2015). While some evidence suggests a potential enrichment in quiescent HSCs in these arteriolar domains, these findings remain controversial to date.

1.3 Downstream the Hematopoietic Hierarchy; Progenitor Cell Niches

Moving beyond HSCs, a number of recent studies have attempted to dissect the spatial distribution and functional interactions that underlie the survival and fate of multipotent progenitors (MPPs) and lineage-committed progenitors (Wei and Frenette 2018). Two fundamental challenges compound the characterization of such niches: first, the lack of faithful reporters to unambiguously visualize cells transitioning

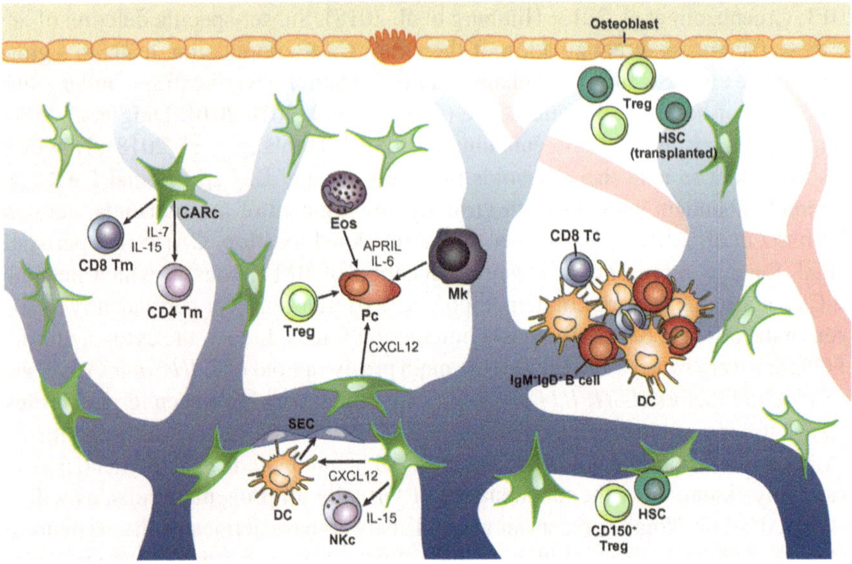

Fig. 2 Graphical depiction of anatomical niches and cellular interactions of mature hematopoietic cells in the BM. From left to right: $CD8^+$ and $CD4^+$ memory T cells display no specific global localization patterns, but have been observed to interact with CARc and depend on IL-7 and IL-15 production by this cell type. A specific subtype of cDCs is perisinusoidal and regulates SEC proliferation and vascular remodeling. Plasmacytoid DCs and NK cells depend on CARc-derived CXCL12 and IL-15, respectively. PCs have been described to specifically and independently interact with eosinophils (Eos), megakaryocytes and T_{regs} and depend on soluble growth factors such as APRIL and IL-6. In transplantation settings, T_{regs} and donor HSCs have been visualized in the proximity of endosteal surfaces (top). A distinct subset of $CD150^+$ Tregs has been proposed to regulate the HSC pool through close spatial interactions (bottom). Clusters of DCs, T cells and mature IgM^+IgD^+ B cells recognizing blood-borne antigens have been described in the proximity of BM microvessels (middle)

through discrete intermediate stages of differentiation without the use of complex panels of cell surface markers, which often do not perform well in 2D or 3D imaging techniques (Gomariz et al. 2019); second, the fact that subtle changes upstream and downstream in the hematopoietic hierarchy are rapidly propagated to progenitor cells. Thus, dissecting whether niche alterations directly impinge on a distinct progenitor compartment via direct mechanisms, or do so by perturbing the homeostatic balance of the hematopoietic continuum, may be often challenging. Hence, so far little is known on the potential topographical differences and interacting patterns of (MPPs). Most importantly, as for HSCs, almost invariably, the CARc pool has been shown to interact with, and support the maintenance of most hematopoietic progenitor populations in the BM (Omatsu et al. 2010).

Myelomonocytic progenitors: Whether specific spatial trajectories and sequential occupancy of niches dictate progression along the myeloid lineage has not

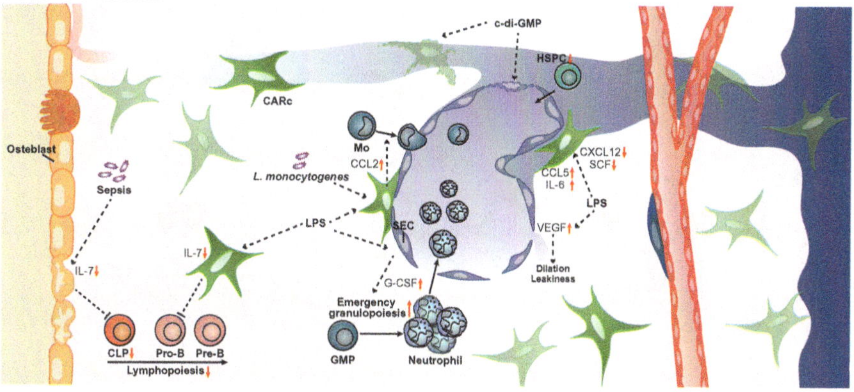

Fig. 3 Schematic depiction of principal alterations of the BM microenvironmental organization during bacterial infections. Left to right: bone-lining osteoblasts are ablated in experimental models of sepsis in a G-CSF-dependent manner. Loss of osteoblasts correlates with reductions of CLPs and leads to lymphopenic states. Impaired B cell differentiation potential due to downregulation of IL-7 production in CARc is also observed in the BM after administration of LPS as well as other inflammatory agents. Stimulation of TLR pathways by LPS or bacterial-derived products in the context of infections with *L. monocytogenes* induces upregulation of CCL2 in perivascular CARc, attracting CCR2-expressing monocytes to perisinusoidal spaces and thereby facilitating their mobilization into the peripheral circulation. TLR4 ligation by LPS additionally induces G-CSF expression in BM ECs, which promotes emergency granulopoiesis in the context of bacterial infections with *E. coli*. Administration of LPS has also been found to originate pronounced and transient sinusoidal vasodilation and increased permeability, through local increases in the production of VEGF in the BM. Finally, prominent upregulation of inflammatory cytokines and downregulation of key matrisomic genes and pro-hematopoietic factors such as SCF and CXCL12 are observed in BM CARc and SECs from mice treated with LPS. Such expression programs may alter the maintenance of HSC subsets and promote their mobilization to extramedullary sites. The bacterial second messenger c-di-GMP has been shown to induce reductions in SECs and CARc by yet undefined mechanisms, potentially perturbing functional homeostatic HSC niches

been studied in depth. Imaging approaches have enabled tracking of granulocyte/macrophage progenitors (GMP), which are immediately downstream of CMP, and are common precursors for neutrophils and monocytes. GMPs were found scattered throughout the BM in steady-state conditions (Hérault 2017). This pattern is perturbed during regenerative responses post-chemotherapy or in the context of malignancies, when GMPs regroup into highly proliferative clusters, which are needed to rapidly replenish the peripheral myeloid compartment. These studies crucially demonstrated that the spatial arrangement of myeloid progenitors is pliable and highly responsive to external forces (Hérault 2017). Moreover, depletion of the entire CARc population leads to the virtual disappearance of GMPs from the BM, which indicates that most likely, GMPs require molecular cues and direct interactions with at least a subset of CARc (Omatsu et al. 2010; Comazzetto et al. 2019). Nonetheless, whether loss of GMPs in this context is due to the disruptive effects of CARc ablation on the upstream HSC and MPP compartments is unknown.

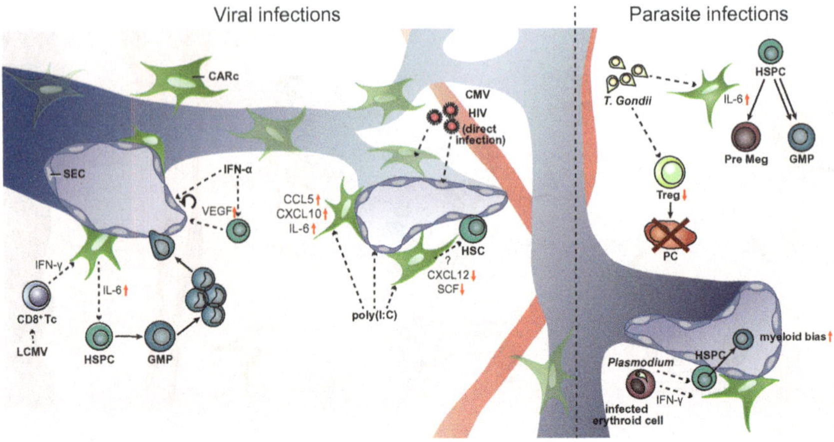

Fig. 4 Effects of viral and parasitic infections and ensuing inflammation in the BM microenvironment. Schematic depiction of the main changes in cellular crosstalk described so far during viral challenge or administration of p:IC, as a viral mimicking agent (left). During acute infections with LCMV, activated $CD8^+$ T cells accumulate in the BM and secrete high levels of IFNγ, which stimulates IL-6 production by CARc. Activation of HSPCs by CARc-derived IL-6 induces differentiation bias toward myelomonocytic cell production. Administration of p:IC or IFNα has been shown to trigger multiple transcriptomic, functional and structural changes in SECs. IFNα enhances hematopoietic cell secretion of VEGF, thereby promoting SEC stimulation and vascular remodeling. Transcriptomic alterations elicited by in vivo administration of pI:C include down-regulation of HSC supportive factors, and prominent upregulation of pro-inflammatory cytokines. Certain viruses, such as HIV and CMV, have also been shown to directly infect mesenchymal and endothelial stromal cells impairing their ability to support hematopoietic differentiation. On the right, a graphical view is shown on the effects of parasitic infections in the stromal infrastructure and hematopoietic and immune cell niches. Infections with the obligate intracellular parasite, *T. gondii*, lead to upregulation of IL-6 in CARc, which in turn promotes biased myeloid cell differentiation from HSCs and concomitant decreases in erythroid cell production. In addition, *T. gondii* has been observed to induce severe depletion of BM T_{regs}, which contributes to a loss of PCs in these tissues. Bottom: murine models of malaria infections with *Plasmodium* parasites elicit major alterations and mobilization of the BM HSPC compartment, which mediate the local effect of IFNγ in the hematopoietic and stromal compartments

Lymphoid progenitors: Although results remain fragmentary and a complete map of the topography underlying B cell development is missing, considerable knowledge on BM lymphopoietic niches has been generated in the past decade. A number of studies suggested the existence of a functional interplay between mature, bone-lining osteoblasts and the earliest stages of lymphoid differentiation. In general, lymphoid progenitors display a spatial bias toward inner bone surfaces, where mature osteoblasts reside. A substantial fraction (~30%) of common lymphoid progenitor cells (CLPs) localize in the proximity of the endosteum (Ding and Morrison 2013). A similar spatial trend has been observed for early pro-B cells, which preferentially accumulate in endosteal regions in both metaphysis and diaphysis and gradually decline toward central BM zones (Park et al. 2013). In line with these findings, osteoblasts are supportive of B cell development in vitro, while in vivo ablation or genetic manipulation of mature osteoblasts induces sharp declines in the number of the earliest CLP in BM (Zhu et al. 2007; Yu et al. 2015; Terashima et al. 2016). Depending on the genetic tools used for osteoblast depletion, downstream B cell progenitor stages, as well as BM-resident, thymus-seeding T cell precursors, were also affected to different degrees in these studies, while HSCs and GMPs remained unaltered. A number of osteoblast-derived factors, such as Dll4, CXCL12 and Wntless, have been documented to potentially mediate the direct cellular crosstalk between lymphoid cells and osteoblasts (Ding and Morrison 2013; Yu et al. 2015; Cao et al. 2015).

Despite the evidence discussed above, the exact contribution of terminally differentiated bone lineage cells to BM lymphopoiesis remains controversial, given that the genetic models employed to target osteoblasts in these studies have not been always specific and could also target immature osteolineage cells, which mostly belong to the CARc subset. Indeed, IL-7, which is the principal cytokine governing lymphopoiesis, is mostly expressed by a subset of CARc (Cordeiro Gomes et al. 2016; Balzano et al. 2019). Compelling studies have uncovered that $IL7^{+}$CARc are indispensable to maintain B cell development in the BM, and most progenitor cells including CLP, pre-B and pro-B cells are directly attached to these cells (Cordeiro Gomes et al. 2016; Fistonich et al. 2018). Since $IL7^{+}$CARc are widely spread throughout the entire BM parenchyma, it seems unlikely that B cell differentiation follows strict regional compartmentalization (Cordeiro Gomes et al. 2016). Notably, the terminal stages of B cell maturation in the BM occur inside the wide lumen of sinusoidal vessels, and require interactions with SECs, which are mediated by integrin α4 and cannabinoid receptor 2 (Pereira et al. 2009).

Erythroid and megakaryopoietic development: The highest cellular flux in the BM occurs along the *erythropoietic and megakaryopoietic* lineage pathways. Like all progenitor cell subsets, the earliest megakaryocyte–erythroid progenitors (MEPs) interact with and are strictly dependent on CARc and their production of SCF (Comazzetto et al. 2019). Notably, subsequent terminal erythroid differentiation occurs in defined anatomical units, termed erythroblastic islands (EBI), which consist of several erythroid progenitor cells in diverse stages of maturation, tightly aggregated around a $CD169^{+}$ nurturing macrophage (Bessis 1958). The precise sequential steps and molecular underpinnings of this process have been dissected in detail, mostly using in vitro models (Chasis and Mohandas 2008). However, in vivo depletion of $CD169^{+}$ macrophages results in a strong reduction of erythroblasts in the BM in steady state, which drives severe anemic states upon subsequent stress (Chow et al. 2013; Ramos 2019). These in vivo findings support the critical physiological role of macrophage interactions in EBIs. A comprehensive analysis of the general topography, number and turnover dynamics of erythroblastic islands in BM is lacking to date. One study proposed that through the remodeling of the extracellular matrix operated by macrophages, EBIs migrate toward sinusoidal vessels as maturation progresses, to facilitate erythrocyte egress into circulation (Yokoyama et al. 2003). A similar differentiation-driven CXCL12-dependent relocation pattern from non-vascular to perisinusoidal regions has been suggested to underlie megakaryopoiesis

(Avecilla et al. 2003). Nonetheless, recent detailed imaging analysis clearly demonstrates that megakaryocytes homogeneously populate the BM, and the majority lie in close contact to sinusoidal vessels due to the high density of this vascular network which strongly restricts the extravascular space (Stegner 2017). SECs contribute to the terminal differentiation of Mks, which emit large cytoplasmic projections from which platelets are shed directly across the endothelial wall into the sinusoidal lumen (Junt et al. 2007).

1.4 Immune Cell Niches: Plasma B Cells, Memory T Cells and Regulatory T Cells

The repertoire of marrow-residing mature cells of the innate and adaptive immune systems is impressively broad, and matches in complexity that of secondary lymphoid organs, as it includes mature subsets of the B and T lymphoid lineages, as well as macrophages, neutrophils, several dendritic cell (DC) subsets and Natural Killer (NK) cells.*$IgD^{+}IgM^{+}$ B* cells are a specialized cell subpopulation of lymphocytes, which can be directly activated by blood-borne pathogens in a T cell-independent manner (Cariappa et al. 2005). These cells are maintained, and possibly primed inside BM-specific perivascular niches formed by clusters of a distinct pool of *dendritic cells* (DCs) and T cells, which aggregate around arterioles or sinusoids (Cariappa et al. 2006; Sapoznikov et al. 2008; Zhang et al. 2019). Notably, BM perivascular DCs have been characterized as cDC2 types, with distinct tissue-specific transcriptional profiles, the ability to activate central memory T cells and a striking function in controlling SEC proliferation and vascular permeability (Zhang et al. 2019). Despite the potential functional relevance of these specific niches, their principal features and remodeling during active immunological responses remain poorly understood to date. Notably, the development of other DC subsets, namely plasmacytoid DCs, has been shown to require interactions with CARc (Kohara et al. 2007).

The accumulation in marrow tissues of *long-lived plasma cells (PCs) and memory B cells* has been long appreciated, but the precise mechanisms underlying this tissue tropism and maintenance are not well understood. After being generated in PLNs, plasmablasts migrate to the BM in a CXCL12–CXCR4 dependent manner, where they differentiate into mature plasma cells, which are preserved as highly active antibody-secreting cells that maintain humoral immunity for prolonged periods of time (Khodadadi et al. 2019). Beyond requiring CARc-derived signals for BM retention, PC maintenance in the BM has been functionally linked to eosinophils, megakaryocytes T_{regs}. Megakaryocytes and eosinophils are rich sources of APRIL and IL-6, which are essential cytokines for PC development, and both cell types have been observed in frequent association with PC in BM tissues (Chu et al. 2011; Zehentmeier et al. 2014). Accordingly, *c-mpl*$^{-/-}$ mice, which are deficient in megakaryocytes, exhibit twofold reductions in BM PC numbers in homeostasis, and delayed development of new PCs upon immunization (Winter et al. 2010). While eosinophil deficiency was initially found to cause relocation of PCs to the spleen and decreased PC numbers in the BM, data from two independent studies contradicted these findings and suggested that eosinophils are dispensable for PC maintenance (Chu et al. 2011; Bortnick et al. 2018; Haberland et al. 2018). Finally, recent work demonstrates that T_{reg} depletion correlates with a substantial decrease in BM PCs and pointed to a potential spatial association and functional crosstalk between both subsets (Zaretsky et al. 2017). Altogether, the data suggest that PCs most likely depend on dynamic interactions within a multifunctional and complex niche (Zehentmeier et al. 2014). Much less explored is how PCs themselves may engage in active regulatory roles in niches of other BM cell types. Intriguingly, frequencies of PCs increase in the BM with aging, and recent data suggest that via their enhanced inflammatory profile and the indirect regulation of stromal cell properties, PCs promote aging-dependent myeloid bias of HSCs (Pioli et al. 2019).

Collectively, BM cavities host probably the largest pools of *CD4 and CD8 memory T cells*, which for the most part are maintained in a quiescent state and provide immunity against systemic challenges (Tokoyoda et al. 2010; Pascutti et al. 2016). However, whether the BM is a sanctuary for the long-term persistence of sessile memory T cells or rather the stopping station where an actively recirculating pool transiently arrests to collect proliferation and survival cues is yet unresolved and a matter of contention (Chang et al. 2018). Irrespective of this, subpopulations of CARc are the main providers of IL-7 and IL-15, which maintain and drive proliferation and therefore control the absolute number of memory T cells (Tikhonova et al. 2019; Helbling et al. 2019; Cui et al. 2014). Other than such dependency and interaction with CARc, no specific spatial patterns have been observed for memory T cells, which appear scattered through the BM (Tokoyoda et al. 2009). Nonetheless, as mentioned above memory T cells are primed by blood-borne cDCs in perivascular regions where they presumably reside (Sapoznikov et al. 2008; Zhang et al. 2019; Cavanagh et al. 2005).

T_{regs} have a crucial role in the homeostasis of the immune system and contribute to self-tolerance (Campbell and Koch 2011). BM-resident T_{regs} represent a major fraction of the entire systemic pool and are imprinted with tissue-specific properties (Zaretsky et al. 2017). In vivo imaging experiments revealed a close association of T_{regs} with transplanted allogenic HSPCs, within the proximity of endosteal surfaces (Fujisaki et al. 2011). However, detailed 3D imaging reveals that in steady state, T_{regs} are found dispersed throughout the entire BM parenchyma (Gomariz et al. 2018), which therefore argues that regional clustering could be circumstantially induced by irradiation and/or transplantation of allogeneic HSCs. Interestingly, two recent studies show that the ablation of T_{regs} in vivo leads to loss of endogenous HSC quiescence, expansion of progenitors, as well as long-term HSCs in the BM. One study suggested that a specific subset of $CD150^{+}T_{regs}$ exerts control of HSC maintenance in a direct fashion through spatial proximity and production of adenosine (Hirata et al. 2018). Alternatively, data from another group showed that T_{reg} depletion disrupts CARc function, leading to a reduction in SCF and IL-7 production, which in turn blocks B lymphoid differentiation (Pierini et al. 2017). These findings would imply

that the subsequent effects observed in the HSC compartment would indirectly derive from the downstream perturbation of B lymphopoiesis.

Collectively, as evidenced from the discussion above, although incomplete maps of cell types and cellular processes in the BM are being gradually drawn from experimental studies, a holistic view of the functional anatomy of the BM is missing.

2 Inflammation-Induced Hematopoiesis

2.1 Emergency Hematopoiesis and BM Suppression

Among the best described BM responses to pathogenic infections and inflammatory stimuli is the reorganization of the hierarchy and rerouting of differentiation trajectories followed by progenitors, in a switch from homeostatic blood cell production, to demand-driven, so-called emergency hematopoiesis (Zaretsky et al. 2013; Manz and Boettcher 2014). Fast changes in BM function are typically geared toward rapidly increasing the supply of distinct hematopoietic lineages to replenish subsets exhausted during pathogen clearance (Zaretsky et al. 2013; Manz and Boettcher 2014). Emergency responses are tailored to the specific nature of the infectious agent, the type of immune responses triggered by them and the preferential inflammatory cytokine profiles emerging in the process. In the majority of experimental models of acute bacterial infections studied, emergency myelopoiesis is favored at the expense of the production of lymphoid cells (Schultze et al. 2019). Myelomonocytic cells are short-lived, they form the primary line of defense against pathogens, and they play fundamental roles in tissue regeneration during the resolution of inflammation. For this, myeloid cells are in high demand in the earliest phases of infections. As an example, during a bacterial challenge, neutrophils are rapidly recruited to peripheral tissues, and therefore, emergency granulopoiesis is the most common effect in such cases, which is similarly triggered by the administration of bacterial mimicking agents, such as Lipopolysaccharide (LPS) (Manz and Boettcher 2014). Nonetheless, certain acute infections with intracellular antigens may instead preferentially skew myelopoiesis toward the production of monocytes, which are deployed to peripheral tissues for in situ differentiation (Schürch et al. 2014; Shi et al. 2011).

Hematopoietic shifts toward myeloid programs are normally accompanied by a failure in lymphoid development in the BM, which is already evident at the earliest stages of lineage commitment and may be triggered via different mechanisms depending on the infection (Ueda 2005). Of note, while multiple pathogens and inflammatory conditions lead to acute activation of hematopoietic programs, others directly cause a general suppression of BM hematopoiesis (Pascutti et al. 2016; Rosenfeld and Young 1991). This is not only the case of viral infections, classically associated with defective BM function, but also of some obligate intracellular bacteria (Smith 2018). Most importantly, while emergency programs can be beneficial and have most likely evolved to anticipate the need for rapid deployment of innate immune cells, inflammatory stimuli can gradually harm BM function, especially during chronic processes. The potential alterations that persistent inflammation causes in hematopoiesis and HSC function have been mostly explored in the context of non-infectious systemic conditions such as atherosclerosis or diabetes, while little is known on the pathophysiological mechanisms driving BM dysfunction in chronic infections (Chavakis et al. 2019).

It is important to note that beyond infections, conditions resulting in profound tissue destruction, massive cell death and potent release of inflammatory factors trigger adaptive stress responses, which may largely resemble emergency programs activated by pathogenic insults. This is the case of myeloablative conditioning regimens such as ionizing radiation and chemotherapeutic cytoreductive drugs, which efficiently deplete immature and mature hematopoietic compartments, inducing severe cytopenic states and eventually triggering enhanced myeloid cell production. Among them, myeloablative drugs such as 5-FU have been widely employed as experimental models to provide key insight on the regulation of stress hematopoiesis. Nonetheless, the molecular mechanisms underlying these processes significantly differ from those elicited during pathogenic infections and will therefore not be discussed in the context of this review.

2.2 *Inflammation-Induced Adaptations in HSPC Function*

Recent studies have attempted to dissect with a great level of molecular detail, how inflammatory factors are integrated by hematopoietic cells and translated into distinct responses within the BM. Major attention has been focused on HSPCs, as sensing agents of inflammation and primary nodes, from which downstream adjustments in terminal hematopoietic output are rapidly orchestrated (reviewed in detail in Chavakis et al. 2019; Pietras 2017). The readjustment of the primitive compartment of the HSC pool is generally characterized by the massive exit from quiescence, priming toward myeloid lineage differentiation and the mobilization from the BM to extramedullary sites. These adaptive changes can be operated via three principal mechanisms: (i) the direct sensing of pathogen-associated molecular products (PAMPs) through expression specific receptors, (ii) the acute action of pro-inflammatory cytokines locally or systemically produced by immune and stromal cells, and (iii) the direct infection by intracellular pathogens.

HSCs express different Toll-like receptors (TLRs), which render them susceptible to a repertoire of PAMPs (Nagai et al. 2006; Liu et al. 2015). Studies using LPS administration have shown that direct activation of TLR4 induces HSC priming toward myeloid differentiation and/or abnormal proliferation (Zhang et al. 2016; Takizawa et al. 2011). Of note, when repeated or prolonged in time, chronic TLR4 stimulation results in a loss of HSC fitness, which is linked to prolonged proliferative stress (Esplin et al. 2011; Takizawa et al. 2017). LPS also induces the secretion of cytokines by HSPCs, which may have paracrine functions in vivo (Zhao et al. 2014). Nonetheless, TLRs lead to specific and distinct alterations of cellular fate in different progenitor cells (Schmid et al. 2011; Welner et al. 2008), and our understanding of how the combinatorial and simultaneous activation of such pathways at different levels of the hierarchy ultimately alters the outcome during life infections remains scarce.

HSPCs additionally respond to a variety of pro-inflammatory cytokines, which strongly contribute to infection-induced changes in hematopoiesis (Pietras 2017). While some of these factors are released systemically, others are locally produced in the BM microenvironment by stromal cells, as will be discussed in the following section. In general, a common feature of cytokines such as G-CSF, M-CSF, IL-3 and IL-6 is their ability through the specification of myeloid lineage in HSPCs (Chavakis et al. 2019; Pietras 2017). Myeloid fate of HSCs is also driven by Tumor necrosis factor-α (TNF-α) and IL-1β. However, while TNF-α promotes the survival of HSCs and the apoptosis of progenitor cells, IL-β also enhances HSC proliferation (Yamashita and Passegué 2019; Pietras et al. 2016). Of note, chronic stimulation with IL-1β results in a strong, though reversible, decline in HSC self-renewal (Pietras et al. 2016). Activation of HSC cycling is also a hallmark of type I and II interferons (IFN), which are strongly produced during immune responses to almost all pathogens, but especially prominent in the context of infections with viruses and intracellular bacteria (Boxx and Cheng 2016). IFNα signaling via IFNα receptor (IFNAR) transiently forces HSCs out of quiescence, which could contribute to compensate HSC consumption due to increased and accelerated differentiation (Essers 2009). However, chronic stimulation of this pathway can again be detrimental to the HSC pool, leading to pronounced functional deficits and potential exhaustion (Walter et al. 2015). Type II IFNγ has also been linked to enhanced proliferation, accelerated differentiation of a subset of myeloid biased HSCs and eventually exhaustion of self-renewal in the context of infections (Baldridge et al. 2010; Matatall et al. 2016; Matatall et al. 2014). However, the precise effects of IFNγ signaling in HSCs are rather complex and context-dependent, as it has also been shown to suppress HSCs and lineage-committed progenitors or favor monopoiesis in different experimental models (Bruin et al. 2012; Bruin et al. 2014).

Finally, it has been proposed that both the residence in protected niches and the prolonged hibernation endow HSCs with specific molecular mechanisms that renders them highly resistant to infections (Wu 2018). Nonetheless, a number of intracellular agents have been found to display specific tropism and infect primitive hematopoietic subsets, which are highly enriched in HSCs. These include *Mycobacterium tuberculosis*, Human Herpesvirus 7 or Human Cytomegalovirus (CMV) (Simmons et al. 1990; Mirandola et al. 2000; Tornack et al. 2017). The potential of HSCs to act as reservoirs for infectious agents has major implications in the context of BM transplantation. However, beyond this fact, the extent to which direct infection impacts the self-renewal or differentiation capacity of HSCs and consequently the ability to contribute to hematopoiesis of HSCs remains mostly unknown, except for certain viruses. For instance, Parvovirus B19, which targets erythroid progenitors, arrests differentiation and leads to red cell aplasia (Brown and Young 1996). Therefore, it is conceivable that direct infection could have important functional consequences on HSC potential.

2.3 *Trained Immunity*

Immunological memory has long been recognized as a fundamental yet exclusive feature of the adaptive immune system. Nonetheless, recent studies have uncovered that during pathogenic infections, innate immune subsets, such as myeloid or NK cells, undergo functional adaptations, which allow them to more efficiently respond to subsequent challenges (O'Sullivan et al. 2015; Quintin et al. 2014). When mediated by myeloid cells, this phenomenon, which has been termed trained immunity, is elicited by a number of stimuli, including LPS, the fungal polysaccharide ß-glucan or the Bacillus Calmette Guerin vaccine (Netea et al. 2016). First encounters with these challenges result in functional priming of myeloid cells, which then display enhanced, non-specific responses to secondary events with the same or even other stimuli, conferring cross-protection against unrelated pathogens. Given the limited average half-life and rapid consumption of mature myeloid cells, the medium to long-term persistence of trained immunity relies on the functional imprinting of HSPCs, via activation of epigenetic, transcriptional and metabolic programs, which are passed on to the myeloid progeny for prolonged periods of time (Netea et al. 2020). The BM is likely the primary site for such specific education of HSPCs, which would rely on the provision of key specific factors by cells of the stromal and hematopoietic microenvironment during the primary challenge. For instance, β-glucan-mediated proliferation, pro-myeloid shift and metabolic reprogramming are strongly dependent on IL-1β, for which CARc and ECs have been described as a relevant source during different inflammatory challenges (Helbling et al. 2019; Chavakis et al. 2019; Mitroulis et al. 2018). Therefore, the potential contribution of inflammation-related remodeling of BM stromal subsets to the establishment of trained immunity remains elusive and deserves to be explored.

3 Inflammatory-Induced Remodeling of the BM Microenvironment

How do pathogenic infections and ensuing inflammatory conditions modify the native organization of BM tissues, alter functional relationships and consequently shape hematopoiesis both acutely and chronically? What is becoming evident is that the cells of the stromal compartment, as prime organizers of tissue biology, are also highly sensitive to infections, and their responses have a major impact on organ function.

3.1 Bacterial Infections

One of the most conspicuous effects of inflammation, which is not exclusive to BM tissues, but a generalized response of various vascular beds, is the rapid dilation of microvessels. Indeed, administration of LPS causes substantial widening of BM intrasinusoidal volumes, and triggers an angiogenic response through the increased proliferation of SECs (Vandoorne et al. 2018). Detailed structural analyses have shown that vascular remodeling correlates with abnormal leakage caused by loosening of the intracellular junctions in the endothelium. These effects are at least partially mediated by increases in systemic levels of VEGF, and facilitate increased nutrient exchange, sensing of systemic signals and cellular trafficking in and out of the marrow (Vandoorne et al. 2018). The loss of barrier function is thought to contribute to inflammation-induced increases in BM cellular output, mobilization of HSPCs, and to the emergence of transient anemia, which typically accompanies emergency responses (Lim et al. 2014). Physical and morphological changes in sinusoids may be linked to the strong alterations in transcriptomic profiles observed in SEC subsets after in vivo LPS challenge. These changes preferentially affect matrisomic genes, integrins, adhesion molecules, as well as chemokines, which could influence the ability of SECs to preserve vascular integrity upon inflammatory signaling (Helbling et al. 2019). LPS sensing also modifies the expression profile of angiocrine factors secreted by both SECs and AECs, thereby modulating their potential to influence stress-induced hematopoiesis (Helbling et al. 2019). For instance, ECs of multiple tissues, including BM, critically promote emergency granulopoiesis by secreting G-CSF, early after LPS administration or infection with Escherichia coli. G-CSF secretion is directly triggered by the activation of the TLR4-Myd88 signaling axis in ECs, and induces myeloid bias, enhanced proliferation and neutrophil output in this context (Boettcher et al. 2014).

Although detailed knowledge is lacking, recent studies demonstrate that perisinusoidal CARc are also highly reactive to bacterial-derived products and inflammatory cytokines. In vivo administration of LPS, TNF-α or G-CSF leads to transient, yet strong, decreases in BM levels of some of the most important CARc-derived factors, such as SCF, CXCL12 and IL-7 (Helbling et al. 2019; Ueda 2005; Day et al. 2015). Similar effects have been observed in the context of life infections with *Anaplasma phagocytophilum* (Johns and Borjesson 2012). Detailed transcriptomic analyses have uncovered that in vivo LPS challenge triggers the activation of pathological transcriptomic programs in CARc, which are characterized by the downmodulation of genes involved in survival and maintenance of HSPCs (*Cxcl12, Kitl* and *Bmp4*), decreases in the main pro-lymphoid factor *Il7* and the upregulation inflammatory genes with proven capacity to promote myeloid fate (*Il6*) (Helbling et al. 2019). Collectively, these observations strongly suggest that the impact in CARc is a major contributing factor to the suppression of B lymphopoiesis and dysregulation of HSPC maintenance induced by bacterial pathogens. It is, however, unclear to what extent these transcriptional changes are driven by the direct activation of TLRs, or indirectly through the action of inflammatory cytokines secreted during infections. CARc express the receptors for a number of cytokines (TNFR and G-CSFR) and could therefore rapidly react to cytokines locally secreted by innate immune cells in BM tissues.

Notably, CARc also express specific receptors for other less well-studied PAMPs, which may trigger relevant functional adaptations of stromal cells in specific niches. For instance, in vivo administration of bacterial bis-(3'-5')-cyclic dimeric guanosine (c-di-GMP) has been shown to activate HSPCs, while partially depleting CARc and ECs, by binding to the intracellular receptor stimulator of interferon genes (STING) (Kobayashi et al. 2015). Mesenchymal stromal cells additionally express nucleotide-binding oligomerization domain (NOD)-like receptors (NOD), which are activated by peptidoglycans found in both Gram-positive as well as Gram-negative bacteria. Interestingly, basal, homeostatic stimulation of NOD-dependent pathways by microbiota-derived products is essential to preserve the HSC pool in steady state (Iwamura et al. 2017).

Beyond modulating stromal cell functionality, excessive cytokine production can have destructive effects on stromal networks. Using a model of cecal ligation and puncture, Terashima and colleagues demonstrated that generalized polymicrobial infections and sepsis cause a fast and prolonged ablation of osteoblast populations, which correlates with decreased levels of IL-7 and loss of CLPs (Terashima et al. 2016). Block in lymphoid differentiation leads to pronounced T and B cell lymphopenia and immunodeficiency. Of note, the authors found that decreases in osteoblasts were not mediated by TLR4 signaling, but were rather dependent on G-CSF signaling, which had been found to suppress osteoblast numbers and function in a previous study (Terashima et al. 2016; Day et al. 2015). Perhaps most importantly, in this acute infectious setting, the levels of CXCL12 in the BM were unchanged, thereby suggesting that CARc numbers were probably unaffected. These results evidence that strong differences exist in the cellular mechanisms and potential impact in stromal components of distinct pathogenic infections, and thus emphasize the need for individualized studies.

An additional frequent consequence of the functional decline of BM stroma is the weakening of molecular cues that promote BM retention of HSPCs, which causes them to mobilize and temporarily reside in extramedullary sites (Johns and Christopher 2012). Exit from BM may serve not only to protect HSCs, by opening facultative niches during BM stress, but also to efficiently deploy a circulating pool of progenitor cells to peripheral tissues, for their on-site differentiation into a pool of active innate immune cells to fight infections (Massberg et al. 2007). Studies using models of acute bacterial challenge with *E. coli* have shown that long-term multilineage HSC mobilization to splenic tissues required the coordinated activation of both TLR4 and NOD receptor signaling in radioresistant cells (Burberry et al. 2014). From this study, the nature of the stromal cell was not clear, however, HSPC emergence in the spleen relied on the decrease of CXCL12 levels in the BM via G-CSF, thereby suggesting a potential involvement of CARc in this process. Notably, beyond the suppression of pro-retention signals, CARc may also actively promote exit from the BM of different cells. Sensing of low levels of circulating LPS prompts CARc to upregulate CCL2, a potent chemokine for CCR2-expressing monocytes. This effect is pivotal for the development of efficient innate immune responses to bacterial agents such as *Listeria monocytogenes*, as it drives monocyte localization to perisinusoidal spaces and facilitates their egress into the systemic circulation (Shi et al. 2011). Of note, progenitor cells and a fraction of HSCs, which get activated during stress responses to myocardial infarction, express CCR2 (Dutta et al. 2015). Therefore, it is possible that CCR2-driven chemoattraction plays a role in the bacterial-induced mobilization of immature progenitor cells (Si et al. 2010).

Finally, stromal orchestration of hematopoiesis may be compromised due to direct infections by bacterial pathogens. In fact, studies in humans suggest that the intracellular bacteria *Mycobacterium tuberculosis* infects a subset of mesenchymal progenitor cells, which express CD271 and could represent a reservoir for dormant, therapy-resistant bacteria (Das et al. 2013). Collectively, it seems clear that most common infections have an impact on the BM microenvironment, which needs to be further dissected in detail for its likely implications in organ function and hematopoiesis.

3.2 Viral Infections

Viral infections have been associated with a number of BM manifestations, which range in severity from transient suppression of hematopoietic output to severe aplastic anemias and organ failure (Pascutti et al. 2016; Rosenfeld and Young 1991). The mechanisms operating these alterations vary depending on the pathogenic agent and the magnitude, duration and type of immune responses. As discussed above, antiviral cytokines, such as IFNs, have pleiotropic and rather complex effects on HSPCs, which dictate the way in which BM function evolves in the course of viral infections (Smith et al. 2016). However, prime contributions to the remodeling of hematopoietic function are driven by perturbations of stromal functionality, which have been less studied to date (Nombela Arrieta and Isringhausen 2016).

Certain viruses have been observed to infect endothelial and mesenchymal populations and suppress their hematopoietic supportive capacity. Latent infections with Human Cytomegalovirus (H-CMV) are highly prevalent and become reactivated in immunosuppressed individuals, causing severe and even fatal complications in patients undergoing allogeneic BM transplantations (Boeckh and Ljungman 2009). In vitro studies suggest that CMV directly infects stromal cell cultures, impairing their ability to support hematopoietic progenitor cell expansion (Simmons et al. 1990). Moreover, CMV infection leads to defective cytokine secretion of mesenchymal cells in vivo, which points to a direct involvement of stromal dysfunction in graft failure and severe myelosuppression induced by CMV in transplantation settings (Mayer et al. 1997; Apperley et al. 1989). Notably, similar findings have been reported for Human Immunodeficiency Virus (HIV), which is frequently associated with morphological BM abnormalities, defective HSPC growth and cytopenic states (Scadden et al. 1989). HIV infects primary endothelial and mesenchymal cells derived from human BM samples, but not HSCs (Moses et al. 1996; Scadden et al. 1990). Infection of mesenchymal stroma leads to a loss in supportive capacity of hematopoietic colony growth, which is reversed after retroviral therapy together with the morphological abnormalities in BM tissues (Isgrò et al. 2005). Mechanistically, two HIV proteins have been implicated in the degradation of MSC activity and survival in vitro (Beaupere et al. 2015). Finally, recent studies have shown that SARS-CoV-2 infects endothelial cells in a number of organs, and that this tropism could potentially underlie the fatal consequences of this disease (Teuwen et al. 2020). It is therefore tempting to speculate that direct targeting of BM endothelial cells could contribute to the hematopoietic syndromes that have been observed in patients undergoing SARS-CoV-2 infections.

Beyond the direct cytopathic effects of viruses in stromal cells, BM microenvironmental defects may emerge from the effects of antiviral cytokines in stromal components. Our group found that in vivo administration of polyinosinic–polycytydilic acid (p:IC) injection, used as a surrogate of viral-induced inflammation because of its ability to induce massive IFN production, led to very pronounced changes in global gene expression programs of SECs, AECs, CARc and PαS. The majority of the transcriptomic changes elicited were shared with those observed after LPS treatment, thereby suggesting the existence of a stereotypic response to inflammatory states that could engage stroma in the deployment of emergency hematopoietic programs (Helbling et al. 2019). In the case of CARc, these included downregulation of HSPC supportive factors and increased pro-myeloid cytokines that could aid in the rapid differentiation along this lineage. In line with these findings, during life acute infections with the widely employed *model Lymphocytic Choriomeningitis Virus* (LCMV) infection, sensing of IFNγ by CARc induces their production of IL-6, which promotes lineage bias in HSPCs, thus accelerating reactive myelomonopoiesis (Schürch et al. 2014). Dysregulation of the local cytokine environment and upregulation of an inflammatory milieu has also been observed after infection with murine CMV (M-CMV), although the consequences in stromal functionality were not evaluated in this model (Hirche et al. 2017). Finally, IFNs have also been observed to indirectly target BM vascular networks at structural and functional levels. p:IC-induced IFNα leads to the upregulation of vascular endothelial growth factor (VEGF) expression by hematopoietic cells, which in turn promotes EC proliferation, vascular remodeling and enhanced vascular support of HSC cycling (Prendergast et al. 2017). It is important to note that, despite being best characterized in the context of antiviral responses, type I IFNs are also highly produced upon infections with intracellular bacteria (Smith 2018; MacNamara et al. 2011). Therefore, many of the described effects of this family of cytokines both in the hematopoietic and non-hematopoietic compartments may be recapitulated during infections with these intracellular pathogens.

3.3 Parasite Infections

Major effects on hematopoiesis and substantial readjustment of stromal cell functions have also been observed in the context of parasitic infections. Similar to what was described for LCMV, the intracellular parasite *Toxoplasma gondii* causes transient BM hypoplasia through blockage of early erythropoietic progenitor differentiation and induction of GMP accumulation and a shift toward myelomonopoiesis. These effects were found to depend on IL-6 production by a population of BM mesenchymal stromal cells, which mostly corresponded to CARc, based on the cell surface phenotype (Chou et al. 2012). Interestingly, infections with *T. gondii* have also been shown to perturb the homeostasis of the immune microenvironment, leading to the transient disappearance of T_{regs}, which has been causally linked to the subsequent clearance of PCs from the BM (Zaretsky et al. 2017). These findings clearly illustrate the need to understand how repeated pathogenic infections may undermine the maintenance of immunological memory via the erosion of its protective tissue microenvironment, thereby shaping future responses to pathogenic challenges.

Malaria, which is caused by different species of *Plasmodium*, is arguably among the parasitic infections creating a larger global health burden, and its effects in hematopoiesis and BM morphology have long been appreciated (Silverman et al. 1987). BM tissues are a major reservoir for the replication of Plasmodium, which infect and induce functional changes in erythroid progenitors and cause severe anemias (Venugopal et al. 2020). Studies using murine models of malaria uncovered an intense phenotypic and functional restructuring of MPP subsets upon infection, which was mediated by the effect of high levels of IFNγ (Belyaev 2010; Vainieri et al. 2016). In fact, IFNγ signaling also triggered CCR2-dependent mobilization of myeloid progenitors to the periphery and mostly splenic parenchyma, which was important for the resolution of the infection (Belyaev et al. 2013). Interestingly, both of these effects recapitulate the reported modifications observed and discussed in the context of viral and/or intracellular bacterial infections, where IFNγ acts on stromal cell function, most likely CARc, to mediate changes in HSPC functionality and trafficking. This may reflect the existence of a stereotypic response to support emergency output, while also facilitating the rapid supply of undifferentiated cells to peripheral tissues for in situ maturation and immune activation.

4 Concluding Remarks

Our knowledge of the complex topography of hematopoietic and immune cellular processes in the BM has increased exponentially in the last years. In this context, the stromal compartment emerges as a powerful orchestrator of BM dynamics. Stromal components not only provide a structural skeleton but also actively engage in the direct communication and control of hematopoietic cell fate decisions. It is now clear that throughout the entire lifespan, stromal infrastructures are malleable and undergo adaptations, which modulate BM function in response to demand. Stromal networks also undergo destruction and functional degradation in the context of pathological conditions, such as those induced by microbial infections. Finally, distinct stromal components are major players in the process of tissue regeneration post injury. This structural and functional plasticity is thus crucial to the preservation of tissue homeostasis.

Importantly, many cardinal questions remain open and warrant exciting investigations in the near future. For instance, how do chronic, sustained or repeated exposures to pathogenic infections gradually interfere with the regenerative capacity of BM stromal cells? Are both the tissue microarchitecture and functionality of stromal cells fully regained after a major challenge, or may defective restoration of a homeostatic microenvironment result in the gradual erosion of BM functionality? It will be important to determine if potential tissue scarring is linked to the emergence of pro-inflammatory signatures in stromal cells, and whether this renders the BM prone to aging-related pathological traits, clonal hematopoiesis or hematological neoplasias. Other aspects of which we know little to date pertain to the effect of infections and/or inflammatory processes in the ability of BM tissues to host immune cell activation, promote humoral responses and maintain immunological memory. While most studies have focused on the effects on progenitor cell niches, whether infections lead to global redistribution of memory T cells and plasma cells toward distant tissues, and how this may affect recall responses and vaccine development, remains to be studied. These and other relevant questions will be best addressed through the widespread and routine application of single-cell profiling and multidimensional imaging, which offer a more granular view on the specific compositional and structural reconfiguration of the BM microenvironment during disease.

References

Acar M et al (2015) Deep imaging of bone marrow shows non-dividing stem cells are mainly perisinusoidal. Nature 526:126–130

Apperley JF et al (1989) The effect of cytomegalovirus on hemopoiesis: in vitro evidence for selective infection of marrow stromal cells. Exp Hematol 17:38–45

Avecilla ST et al (2003) Chemokine-mediated interaction of hematopoietic progenitors with the bone marrow vascular niche is required for thrombopoiesis. Nat Med 10:64–71

Baccin C et al (2019) Combined single-cell and spatial transcriptomics reveal the molecular, cellular and spatial bone marrow niche organization. Nat Cell Biol 22:1–30

Baldridge MT, King KY, Boles NC, Weksberg DC, Goodell MA (2010) Quiescent haematopoietic stem cells are activated by IFN. Nature 465:793–797

Balzano M et al (2019) Nidogen-1 contributes to the interaction network involved in Pro-B cell retention in the peri- sinusoidal hematopoietic stem cell niche. Cell Rep 26:3257-3271.e8

Baryawno N et al. (2019) A cellular taxonomy of the bone marrow stroma in homeostasis and leukemia. Cell:1–35. https://doi.org/10.1016/j.cell.2019.04.040

Beaupere C et al (2015) The HIV proteins Tat and Nef promote human bone marrow mesenchymal stem cell senescence and alter osteoblastic differentiation. Aging Cell 14:534–546

Belyaev NN et al (2010) Induction of an IL7-R+c-Kithi myelolymphoid progenitor critically dependent on IFN-γ signaling during acute malaria. Nat Immunol:1–10. https://doi.org/10.1038/ni.1869

Belyaev NN, Biró J, Langhorne J, Potocnik AJ (2013) Extramedullary myelopoiesis in malaria depends on mobilization of myeloid-restricted progenitors by IFN-γ induced chemokines. PLoS Pathog 9:e1003406

Bessis M (1958) [Erythroblastic island, functional unity of bone marrow]. Rev Hematol 13:8–11
Boeckh M, Ljungman P (2009) How we treat cytomegalovirus in hematopoietic cell transplant recipients. Blood. https://doi.org/10.1182/blood
Boettcher S et al (2014) Endothelial cells translate pathogen signals into G-CSF-driven emergency granulopoiesis. Blood 124:1393–1403
Bortnick A, Chernova I, Spencer SP, Allman D (2018) No strict requirement for eosinophils for bone marrow plasma cell survival. Eur J Immunol 48:815–821
Boxx GM, Cheng G (2016) The Roles of Type I Interferon in Bacterial Infection. Cell Host Microbe 19:760–769
Brown KE, Young NS (1996) Parvoviruses and bone marrow failure. Stem Cells 14:151–163
Bruns I et al (2014) Megakaryocytes regulate hematopoietic stem cell quiescence through CXCL4 secretion. Nat Med 20:1315–1320
Burberry A et al (2014) Infection mobilizes hematopoietic stem cells through cooperative NOD-like receptorand toll-like receptor signaling. Cell Host Microbe 15:779–791
Campbell DJ, Koch MA (2011) Phenotypical and functional specialization of FOXP3+ regulatory T cells. Nat Rev Immunol 11:119–130
Cao J et al (2015) Ablation of wntlessin endosteal niches impairs lymphopoiesis rather than HSCs maintenance. Eur J Immunol 45:2650–2660
Cariappa A et al (2005) Perisinusoidal B cells in the bone marrow participate in T-independent responses to blood-borne microbes. Immunity 23:397–407
Cariappa A, Chase C, Liu H, Russell P, Pillai S (2006) Naive recirculating B cells mature simultaneously in the spleen and bone marrow. Blood 109:2339–2345
Cavanagh LL et al (2005) Activation of bone marrow–resident memory T cells by circulating, antigen-bearing dendritic cells. Nat Immunol 6:1029–1037
Chang H-D, Tokoyoda K, Radbruch A (2018) Immunological memories of the bone marrow. Immunol Rev 283:86–98
Chasis JA, Mohandas N (2008) Erythroblastic islands: niches for erythropoiesis. Blood 112:470–478
Chavakis T, Mitroulis I, Hajishengallis G (2019) Hematopoietic progenitor cells as integrative hubs for adaptation to and fine-tuning of inflammation. Nat Immunol 20:1–10
Chou DB et al (2012) Stromal-derived IL-6 alters the balance of myeloerythroid progenitors during Toxoplasma gondii infection. J Leukoc Biol 92:123–131
Chow A et al (2013) CD169+ macrophages provide a niche promoting erythropoiesis under homeostasis and stress. Nat Med 19:429–436
Chu VT et al (2011) Eosinophils are required for the maintenance of plasma cells in the bone marrow. Nat Immunol 12:151–159
Comazzetto S et al (2019) Restricted hematopoietic progenitors and erythropoiesis require SCF from leptin receptor+ niche cells in the bone marrow. Cell Stem Cell 24:477-486.e6
Cordeiro Gomes A et al (2016) Hematopoietic stem cell niches produce lineage-instructive signals to control multipotent progenitor differentiation. Immunity 45:1219–1231
Crane GM, Jeffery E, Morrison SJ (2017) Adult haematopoietic stem cell niches. Nat Rev Immunol 4:7
Cui G et al (2014) Characterization of the IL-15 niche in primary and secondary lymphoid organs in vivo. Proc Natl Acad Sci 111:1915–1920
Das B et al (2013) CD271+ Bone marrow mesenchymal stem cells may provide a niche for dormant mycobacterium tuberculosis. Sci Transl Med 5:170ra13–170ra13
Day RB, Bhattacharya D, Nagasawa T, Link DC (2015) Granulocyte colony-stimulating factor reprograms bone marrow stromal cells to actively suppress B lymphopoiesis in mice. Blood. https://doi.org/10.1182/blood-2015-02
de Bruin AM et al (2012) IFN induces monopoiesis and inhibits neutrophil development during inflammation. Blood 119:1543–1554
de Bruin AM, Voermans C, Nolte MA (2014) Impact of interferon-γ on hematopoiesis. Blood 124:2479–2486

Ding L, Morrison SJ (2013) Haematopoietic stem cells and early lymphoid progenitors occupy distinct bone marrow niches. Nature 495:231–235

Ding L, Saunders TL, Enikolopov G, Morrison SJ (2012) Endothelial and perivascular cells maintain haematopoietic stem cells. Nature 481:457–462

Dutta P et al (2015) Myocardial infarction activates CCR2(+) hematopoietic stem and progenitor cells. Cell Stem Cell 16:477–487

Esplin BL et al (2011) Chronic exposure to a TLR ligand injures hematopoietic stem cells. J Immunol 186:5367–5375

Essers MAG et al (2009) IFNα activates dormant haematopoietic stem cells in vivo. Nature:1–6. https://doi.org/10.1038/nature07815

Feuerer M et al (2003) Bone marrow as a priming site for T-cell responses to blood-borne antigen. Nat Med 9:1151–1157

Fistonich C et al (2018) Cell circuits between B cell progenitors and IL-7+ mesenchymal progenitor cells control B cell development. J Exp Med 215:2586–2599

Fujisaki J et al (2011) In vivo imaging of Treg cells providing immune privilege to the haematopoietic stem-cell niche. Nature 474:216-219

Gomariz Á et al (2018) Quantitative spatial analysis of haematopoiesis-regulating stromal cells in the bone marrow microenvironment by 3D microscopy. Nat Commun 9:407–415

Gomariz Á, Isringhausen S, Helbling PM, Nombela Arrieta C (2019) Imaging and spatial analysis of hematopoietic stem cell niches. Annals New York Acad Sci 94:284–12

Greenbaum A et al (2013) CXCL12 in early mesenchymal progenitors is required for haematopoietic stem-cell maintenance. Nature 495:227–230

Haberland K et al (2018) Eosinophils are not essential for maintenance of murine plasma cells in the bone marrow. Eur J Immunol 48:822–828

Helbling PM et al (2019) Global transcriptomic profiling of the bone marrow stromal microenvironment during postnatal development, aging, and inflammation. Cell Rep 29:3313-3330.e4

Hérault A et al. (2017) Myeloid progenitor cluster formation drives emergency and leukaemic myelopoiesis. Nature:1–19 (2017). https://doi.org/10.1038/nature21693

Himburg HA et al (2018) Distinct bone marrow sources of pleiotrophin control hematopoietic stem cell maintenance and regeneration. Cell Stem Cell 23:370-381.e5

Hirata Y et al (2018) CD150high Bone Marrow Tregs Maintain Hematopoietic Stem Cell Quiescence and Immune Privilege via Adenosine. Cell Stem Cell 22:445-453.e5

Hirche C et al (2017) Systemic virus infections differentially modulate cell cycle state and functionality of long-term hematopoietic stem cells in vivo. Cell Rep 19:2345–2356

Isgrò A et al (2005) Immunodysregulation of HIV disease at bone marrow level 4:486–490

Itkin T et al (2016) Distinct bone marrow blood vessels differentially regulate haematopoiesis. Nature 532:323–328

Iwamura C, Bouladoux N, Belkaid Y, Sher A, Jankovic D (2017) Sensing of the microbiota by NOD1 in mesenchymal stromal cells regulates murine hematopoiesis. Blood 129:171–176

Johns J, Borjesson D (2012) Downregulation of CXCL12 signaling and altered hematopoietic stem and progenitor cell trafficking in a murine model of acute Anaplasma phagocytophilum infection. Innate Immun 18:418–428

Johns JL, Christopher MM (2012) Extramedullary hematopoiesis: a new look at the underlying stem cell niche, theories of development, and occurrence in animals. Vet Pathol 49:508–523

Junt T et al (2007) Dynamic visualization of thrombopoiesis within bone marrow. Science 317:1767–1770

Junt T, Scandella E, Ludewig B (2008) Form follows function: lymphoid tissue microarchitecture in antimicrobial immune defence. Nat Rev Immunol 8:764–775

Kfoury Y, Scadden DT (2015) Mesenchymal cell contributions to the stem cell niche. Cell Stem Cell 16:239–253

Khodadadi L, Cheng Q, Radbruch A, Hiepe F (2019) The maintenance of memory plasma cells. Front Immunol 10:64–17

Kiel MJ, Iwashita T, Yilmaz ÖH, Morrison SJ (2005) Spatial differences in hematopoiesis but not in stem cells indicate a lack of regional patterning in definitive hematopoietic stem cells. Dev Biol 283:29–39
Kobayashi H et al (2015) Bacterial c-di-GMP affects hematopoietic Stem/progenitors and their niches through STING. Cell Rep 11:71–84
Kohara H et al (2007) Development of plasmacytoid dendritic cells in bone marrow stromal cell niches requires CXCL12-CXCR4 chemokine signaling. Blood 110:4153–4160
Kunisaki Y et al (2013) Arteriolar niches maintain haematopoietic stem cell quiescence. Nature 502:637–643
Lassailly F, Foster K, Lopez-Onieva L, Currie E, Bonnet D (2013) Multimodal imaging reveals structural and functional heterogeneity in different bone marrow compartments: functional implications on hematopoietic stem cells. Blood 122:1730–1740
Lim S et al (2014) VEGFR2-mediated vascular dilation as a mechanism of VEGF-induced anemia and bone marrow cell mobilization. Cell Rep 9:569–580
Liu A et al (2015) Cutting edge: hematopoietic stem cell expansion and common lymphoid progenitor depletion require hematopoietic-derived, cell-autonomous TLR4 in a model of chronic endotoxin. J Immunol 195:2524–2528
MacNamara KC et al (2011) Infection-induced myelopoiesis during intracellular bacterial infection is critically dependent upon IFN- signaling. J Immunol 186:1032–1043
Manz MG, Boettcher S (2014) Emergency granulopoiesis. Nat Rev Immunol 14:302–314
Massberg S et al (2007) Immunosurveillance by hematopoietic progenitor cells trafficking through blood, lymph, and peripheral tissues. Cell 131:994–1008
Matatall KA, Shen C-C, Challen GA, King KY (2014) Type II interferon promotes differentiation of myeloid-biased hematopoietic stem cells. Stem Cells 32:3023–3030
Matatall KA et al (2016) Chronic Infection Depletes Hematopoietic Stem Cells through Stress-Induced Terminal Differentiation. Cell Rep 17:2584–2595
Matsushita Y et al (2020) A Wnt-mediated transformation of the bone marrow stromal cell identity orchestrates skeletal regeneration. Nat Commun 11:1–17
Mayer A et al (1997) Bone marrow failure by cytomegalovirus is associated with an in vivo deficiency in the expression of essential stromal hemopoietin genes. J Virol 71:4589–4598
Méndez-Ferrer S et al (2020) Bone marrow niches in haematological malignancies. Nat Rev Cancer 20:1–14
Mercier FE, Ragu C, Scadden DT (2012) The bone marrow at the crossroads of blood and immunity. Nat Rev Immunol 12:49–60
Mirandola P et al (2000) Infection of CD34(+) hematopoietic progenitor cells by human herpesvirus 7 (HHV-7). Blood 96:126–131
Mitroulis I et al (2018) Modulation of Myelopoiesis Progenitors Is an Integral Component of Trained Immunity. Cell 172:147-155.e12
Morrison SJ, Scadden DT (2014) The bone marrow niche for haematopoietic stem cells. Nature 505:327–334
Moses AV et al (1996) Human immunodeficiency virus infection of bone marrow endothelium reduces induction of stromal hematopoietic growth factors. Blood 87:919–925
Mueller SN, Germain RN (2009) Stromal cell contributions to the homeostasis and functionality of the immune system. Nat Rev Immunol 9:618–629
Nagai Y et al (2006) Toll-like receptors on hematopoietic progenitor cells stimulate innate immune system replenishment. Immunity 24:801–812
Netea MG et al (2020) Defining trained immunity and its role in health and disease. Nat Rev Immunol 173:1–14
Netea MG et al (2016) Trained immunity: A program of innate immune memory in health and disease. Science 352:aaf1098–aaf1098
Nombela Arrieta C, Isringhausen S (2016) The role of the bone marrow stromal compartment in the hematopoietic response to microbial infections. Front Immunol 7:689

Nombela-Arrieta C, Manz MG (2017) Quantification and three-dimensional microanatomical organization of the bone marrow. Blood Adv
O'Sullivan TE, Sun JC, Lanier LL (2015) Natural killer cell memory. Immunity 43:634–645
Omatsu Y et al (2010) The essential functions of adipo-osteogenic progenitors as the hematopoietic stem and progenitor cell niche. Immunity 33:387–399
Omatsu Y, Seike M, Sugiyama T, Kume T, Nagasawa T (2014) Foxc1 is a critical regulator of haematopoietic stem/ progenitor cell niche formation. Nature:1–16. https://doi.org/10.1038/nature13071
Park S-Y et al (2013) Focal adhesion kinase regulates the localization and retention of pro-B cells in bone marrow microenvironments. J Immunol 190:1094–1102
Pascutti MF, Erkelens MN, Nolte MA (2016) Impact of viral infections on hematopoiesis: from beneficial to detrimental effects on bone marrow output. Front Immunol 7:364
Pereira JP, An J, Xu Y, Huang Y, Cyster JG (2009) Cannabinoid receptor 2 mediates the retention of immature B cells in bone marrow sinusoids. Nat Immunol 10:403–411
Pierini A et al (2017) Foxp3^{+} regulatory T cells maintain the bone marrow microenvironment for B cell lymphopoiesis. Nat Commun 8:15068–15113
Pietras EM (2017) Inflammation: a key regulator of hematopoietic stem cell fate in health and disease. Blood 130:1693–1698
Pietras EM et al (2016) Chronic interleukin-1 exposure drives haematopoietic stem cells towards precocious myeloid differentiation at the expense of self-renewal. Nat Cell Biol 18:607–618
Pinho S, Frenette PS (2019) Haematopoietic stem cell activity and interactions with the niche. Nat Rev Mol Cell Biol 20:1–18
Pioli PD, Casero D, Montecino-Rodriguez E, Morrison SL, Dorshkind K (2019) Plasma cells are obligate effectors of enhanced myelopoiesis in aging bone marrow. Immunity:1–29. https://doi.org/10.1016/j.immuni.2019.06.006
Prendergast ÁM et al (2017) IFNα-mediated remodeling of endothelial cells in the bone marrow niche. Haematologica 102:445–453
Quintin J, Cheng S-C, van der Meer JWM, Netea MG (2014) Innate immune memory: towards a better understanding of host defense mechanisms. Curr Opin Immunol 29:1–7
Ramos P et al (2019) Macrophages support pathological erythropoiesis in polycythemia vera and β thalassemia. Nat Med:1–11. https://doi.org/10.1038/nm.3126
Rosenfeld SJ, Young NS (1991) Viruses and bone marrow failure. Blood Rev 5:71–77
Sapoznikov A et al (2008) Perivascular clusters of dendritic cells provide critical survival signals to B cells in bone marrow niches. Nat Immunol 9:388–395
Scadden DT, Zon LI, Groopman JE (1989) Pathophysiology and management of HIV-associated hematologic disorders. Blood 74:1455–1463
Scadden DT et al (1990) Human immunodeficiency virus infection of human bone marrow stromal fibroblasts. Blood 76:317–322
Schmid MA, Takizawa H, Baumjohann DR, Saito Y, Manz MG (2011) Bone marrow dendritic cell progenitors sense pathogens via Toll-like receptors and subsequently migrate to inflamed lymph nodes. Blood 118:4829–4840
Schofield R (1978) The relationship between the spleen colony-forming cell and the haemopoietic stem cell. Blood Cells 4:7–25
Schultze JL, Mass E, Schlitzer A (2019) Emerging Principles in Myelopoiesis at Homeostasis and during Infection and Inflammation. Immunity 50:288–301
Schürch CM, Riether C, Ochsenbein AF (2014) Cytotoxic CD8(+) T cells stimulate hematopoietic progenitors by promoting cytokine release from bone marrow mesenchymal stromal cells. Cell Stem Cell. https://doi.org/10.1016/j.stem.2014.01.002
Shi C et al (2011) Bone marrow mesenchymal stem and progenitor cells induce monocyte emigration in response to circulating toll-like receptor ligands. Immunity 34:590–601
Si Y, Tsou C-L, Croft K, Charo IF (2010) CCR2 mediates hematopoietic stem and progenitor cell trafficking to sites of inflammation in mice. J Clin Investig 120:1192–1203

Silverman PH, Schooley JC, Mahlmann LJ (1987) Murine malaria decreases hemopoietic stem cells. Blood 69:408–413
Simmons P, Kaushansky K, Torok-Storb B (1990) Mechanisms of cytomegalovirus-mediated myelosuppression: perturbation of stromal cell function versus direct infection of myeloid cells. Proc Natl Acad Sci USA 87:1386–1390
Sivaraj KK, Adams RH (2016) Blood vessel formation and function in bone. Development 143:2706–2715
Smith JNP, Kanwar VS, MacNamara KC (2016) Hematopoietic stem cell regulation by Type I and II interferons in the pathogenesis of acquired aplastic anemia. Front Immunol 7:569–613
Smith JNP et al (2018) Type I IFNs drive hematopoietic stem and progenitor cell collapse via impaired proliferation and increased RIPK1-dependent cell death during shock-like ehrlichial infection. PLoS Pathog 14:e1007234
Stegner D et al (2017).Thrombopoiesis is spatially regulated by the bone marrow vasculature. Nat Commun:1–11. https://doi.org/10.1038/s41467-017-00201-7
Sugiyama T, Kohara H, Noda M, Nagasawa T (2006) Maintenance of the hematopoietic stem cell pool by CXCL12-CXCR4 chemokine signaling in bone marrow stromal cell niches. Immunity 25:977–988
Takizawa H, Regoes RR, Boddupalli CS, Bonhoeffer S, Manz MG (2011) Dynamic variation in cycling of hematopoietic stem cells in steady state and inflammation. J Exp Med 208:273–284
Takizawa H, Boettcher S, Manz MG (2012) Demand-adapted regulation of early hematopoiesis in infection and inflammation. Blood 119:2991–3002
Takizawa H et al (2017) Pathogen-induced TLR4-TRIF innate immune signaling in hematopoietic stem cells promotes proliferation but reduces competitive fitness. Cell Stem Cell 21:225-240.e5
Terashima A et al (2016) Sepsis-induced osteoblast ablation causes immunodeficiency. Immunity 44:1434–1443
Teuwen LA, Geldhof V, Pasut A, Carmeliet P (2020) COVID-19: the vasculature unleashed. Nat Rev Immunol:1–3. https://doi.org/10.1038/s41577-020-0343-0
Tikhonova AN et al (2019) The bone marrow microenvironment at single-cell resolution. Nature 505:1–28
Tokoyoda K et al (2009) Professional memory CD4+ T lymphocytes preferentially reside and rest in the bone marrow. Immunity 30:721–730
Tokoyoda K, Hauser AE, Nakayama T, Radbruch A (2010) Organization of immunological memory by bone marrow stroma. Nat Rev Immunol 10:193–200
Tornack J et al (2017) Human and mouse hematopoietic stem cells are a depot for dormant mycobacterium tuberculosis. PLoS ONE 12:e0169119-e169218
Ueda Y (2005) Inflammation and the reciprocal production of granulocytes and lymphocytes in bone marrow. J Exp Med 201:1771–1780
Vainieri ML et al (2016) Systematic tracking of altered haematopoiesis during sporozoite-mediated malaria development reveals multiple response points. Open Biol 6:160038–160113
Vandoorne K et al (2018) Imaging the vascular bone marrow niche during inflammatory stress. Circul Res CIRCRESAHA 118:313302. doi:https://doi.org/10.1161/CIRCRESAHA.118.313302
Venugopal K, Hentzschel F, Nas GVX, Marti M (2020) Plasmodium asexual growth and sexual development in the haematopoietic niche of the host. Nat Rev Microbiol 210:1–13
Walter D et al (2015) Exit from dormancy provokes DNA-damage-induced attrition in haematopoietic stem cells. Nature 520:549–552
Wei Q, Frenette PS (2018) Niches for hematopoietic stem cells and their progeny. Immunity 48:632–648
Welner RS et al (2008) Lymphoid precursors are directed to produce dendritic cells as a result of TLR9 ligation during herpes infection. Blood 112:3753–3761
Winter O et al (2010) Megakaryocytes constitute a functional component of a plasma cell niche in the bone marrow. Blood 116:1867–1875
Wolock SL et al (2019) Mapping distinct bone marrow niche populations and their differentiation paths. Cell Rep 28:302-311.e5

Wu X et al (2018) Intrinsic immunity shapes viral resistance of stem cells. Cell:1–42. https://doi.org/10.1016/j.cell.2017.11.018
Xu C et al (2018) Stem cell factor is selectively secreted by arterial endothelial cells in bone marrow. Nat Commun:1–13 (2018).https://doi.org/10.1038/s41467-018-04726-3
Yamashita M, Passegué E (2019) TNF-α coordinates hematopoietic stem cell survival and myeloid regeneration. Cell Stem Cell 25:357-372.e7
Yokoyama T, Etoh T, Kitagawa H, Tsukahara S, Kannan Y (2003) Migration of erythroblastic islands toward the sinusoid as erythroid maturation proceeds in rat bone marrow. J Vet Med Sci 65:449–452
Yu VWC et al (2015) Specific bone cells produce DLL4 to generate thymus-seeding progenitors from bone marrow. J Exp Med 212:759–774
Zaretsky AG, Engiles JB, Hunter CA (2013) Infection-induced changes in hematopoiesis. J Immunol 192:27–33
Zaretsky AG et al (2017) T regulatory cells support plasma cell populations in the bone marrow. Cell Rep 18:1906–1916
Zehentmeier S et al (2014) Static and dynamic components synergize to form a stable survival niche for bone marrow plasma cells. Eur J Immunol 44:2306–2317
Zhang H et al (2016) Sepsis induces hematopoietic stem cell exhaustion and myelosuppression through distinct contributions of TRIF and MYD88. Stem Cell Reports 6:940–956
Zhang J et al (2019) Bone marrow dendritic cells regulate hematopoietic stem/progenitor cell trafficking. J Clin Invest 129:2920–2931
Zhao JL et al (2014) Conversion of danger signals into cytokine signals by hematopoietic stem and progenitor cells for regulation of stress-induced hematopoiesis. Cell Stem Cell 14:445–459
Zhao M et al. (2014) Megakaryocytes maintain homeostatic quiescenceand promote post-injury regeneration of hematopoietic stem cells. Nat Med 1–8. https://doi.org/10.1038/nm.3706
Zhu J et al (2007) Osteoblasts support B-lymphocyte commitment and differentiation from hematopoietic stem cells. Blood 109:3706–3712

The Roles of IL-7 and IL-15 in Niches for Lymphocyte Progenitors and Immune Cells in Lymphoid Organs

Koichi Ikuta, Takahiro Hara, Shinya Abe, Takuma Asahi, Daichi Takami, and Guangwei Cui

Abstract Lymphoid organs consist of immune cells and stromal cells. The stromal cells produce various cytokines that support the development, maintenance, and response of the immune cells. IL-7 and IL-15 are the major cytokines produced by stromal cells and are essential for the development and maintenance of lymphocytes and innate lymphoid cells (ILCs). In addition, IL-7 is indispensable for the organogenesis of lymphoid organs. However, because the amount of these two cytokines is relatively low, it has been difficult to directly detect their expression. Recently, several groups succeeded in establishing IL-7 and IL-15 reporter mouse lines. As expected, IL-7 and IL-15 were detected in mesenchymal stromal cells in the bone marrow and lymph nodes and in epithelial cells in the thymus. Furthermore, IL-7 and IL-15 were differentially expressed in lymphatic endothelial cells and blood endothelial cells, respectively. In addition to their expression, many groups have analyzed the local functions of IL-7 and IL-15 by using cell-type-specific knockout mice. From these experiments, CXCL12-expressing mesenchymal stromal cells were identified as the major niche for early B cell precursors. Single-cell RNA sequencing (scRNA-seq) analysis has revealed different subpopulations of stromal cells in the lymphoid organs, including those that express both IL-7 and IL-15. Future research is still needed to elucidate which stromal cells serve as the niche for the early precursors of ILCs and NK cells in the bone marrow.

K. Ikuta (✉) · T. Hara · S. Abe · T. Asahi · D. Takami · G. Cui
Laboratory of Immune Regulation, Department of Virus Research, Institute for Frontier Life and Medical Sciences, Kyoto University, Kyoto 606-8507, Japan
e-mail: ikuta.koichi.6c@kyoto-u.ac.jp

T. Asahi
Graduate School of Medicine, Kyoto University, Kyoto 606-8501, Japan

D. Takami
Graduate School of Pharmaceutical Sciences, Kyoto University, Kyoto 606-8501, Japan

T. Nagasawa (ed.), *Bone Marrow Niche*, Current Topics in Microbiology and Immunology 434, https://doi.org/10.1007/978-3-030-86016-5_4

1 Introduction

Lymphoid organs consist of immune cells and stromal cells. The stromal cells produce growth factors, cytokines, chemokines, chemical mediators, and adhesion molecules, which all support the development, maintenance, and response of the immune cells. IL-7 and IL-15 are two cytokines produced by stromal cells and are essential for the development and maintenance of lymphocytes and innate lymphoid cells (ILCs). Lymphoid tissue inducer (LTi) cells, a type of ILCs, are involved in the organogenesis of lymphoid organs. Thus, IL-7 and IL-15 are essential for the development and maintenance of the immune system.

However, analyzing the stromal cells that produce IL-7 and IL-15 is not trivial. The quantity of these cytokines is often too small to be detected in lymphoid tissues. To overcome this difficulty, cytokine reporter mouse lines have been developed, but the reporter expression range depends on the mice being transgenic or knock-in models. Therefore, genome-wide expression analysis of single stromal cells, such as single-cell RNA sequencing (scRNA-seq), is also recommended. This approach can dissect cell subpopulations of different sets of gene expressions.

In this review, first, we will overview the concept of the immune microenvironment that produces immune homeostatic cytokines such as IL-7 and IL-15. Next, we will highlight recent findings from experiments using IL-7 and IL-15 reporter mice. Third, we will describe the local function of these cytokines based on experiments using cell-type-specific IL-7 and IL-15 conditional knockout (cKO) mice. Finally, we will explain recent studies using scRNA-seq of stromal cell subpopulations in lymphoid organs.

2 Cytokine-Producing Immune Microenvironment

2.1 *Immune Microenvironment*

Lymphoid organs consist of two kinds of cells: immune cells and stromal cells. The immune cells include T and B lymphocytes, NK cells, dendritic cells (DCs), and macrophages, which play major roles in immune responses. The stromal cells are composed of various types of adherent cells with different functions. They constitute the immune microenvironment, providing a platform for the immune cells to efficiently function. Stromal cells express adhesion molecules that bind to lymphocytes and produce various soluble factors, such as cytokines, chemokines, and chemical mediators. These factors induce, augment, and modulate the distribution and function of the lymphocytes. Among them, IL-7 and IL-15 are the cytokines most essential for the development and maintenance of lymphocytes and lymphoid organs (Kang and Coles 2012; Ma et al. 2006; Robinette et al. 2017).

2.2 *Production of IL-7 and IL-15*

IL-7 and IL-15 belong to the γc cytokine family (Raeber et al. 2018). IL-7 receptor (IL-7R) consists of a unique IL-7R α-chain (IL-7Rα) and the common γ(γc)-chain. The γc-chain is shared with IL-15R, which also contains IL-15Rα- and IL-2/15Rβ-chains. IL-7R is expressed on early T and B cells, mature T cells, and ILCs (Mazzucchelli and Durum 2007). IL-7 plays an essential role in early lymphocyte development and the maintenance of mature T cells (Raeber et al. 2018). In addition, IL-7 is required for the differentiation of ILCs (Diefenbach et al. 2014). Because LTi cells, a type of group 3 ILC (ILC3), also depend on IL-7 for their development, IL-7 is indispensable for the organogenesis of lymphoid organs such as Peyer's patches (Adachi et al. 1998). In contrast to the broader targets of IL-7, IL-15 is more specific to lymphocytes with cytotoxic or innate features. IL-15 supports the development and maintenance of NK cells, NKT cells, memory CD8 T cells, and CD8$\alpha\alpha^+$ intraepithelial lymphocytes (IELs) of the intestines (Ma et al. 2006). IL-15Rα is important for the trans-presentation of the IL-15/IL-15Rα complex, which is essential for IL-15 signaling in many cases (Ma et al. 2006).

Many kinds of stromal cells produce IL-7 and IL-15. It is reported that IL-7 is produced by mesenchymal stromal cells in the bone marrow, thymic epithelial cells, and fibroblastic reticular cells (FRCs) in the lymph nodes (Link et al. 2007; Namen et al. 1988; Sudo et al. 1989). IL-7 is also expressed in epidermal epithelial cells, intestinal epithelial cells (IECs), and hepatocytes (Matsue et al. 1993; Watanabe et al. 1995). IL-15, on the other hand, is produced in the bone marrow, thymus, and lymph nodes, as well as non-lymphoid organs such as the heart, lung, muscle, and kidney (Grabstein et al. 1994). IL-15 is also produced by DCs, monocytes, and macrophages, more so when the cells are activated (Mattei et al. 2001). However, it is difficult to identify the type and distribution of stromal cells that produce IL-7 or IL-15 in lymphoid organs for mainly two reasons: the low amount of these cytokines in the tissues and the unavailability of specific antibodies for effective immunohistochemistry. Thus, many previous studies detected cytokine mRNA levels but did not identify the cytokine-producing stromal cells.

3 Cytokine-Producing Stromal Cells in Bone Marrow and Lymphoid Organs

To identify the type and distribution of stromal cells expressing IL-7 in lymphoid organs, several groups have utilized cytokine reporter mice (Table 1). Three groups reported IL-7 reporter mice harboring bacterial artificial chromosome (BAC) transgenes with fluorescent or surface protein reporter genes controlled by the IL-7 promoter. IL-7–YFP BAC transgenic reporter mice express their signal in TECs, especially at the corticomedullary junction (Alves et al. 2009). Surprisingly, the IL-7/YFP signal was never detected in peripheral lymphoid organs. The fluorescence

Table 1 IL-7 and IL-15 reporter mice

Cytokine reporter mice are summarized with the cytokine-producing cell types

Reporter line	Mouse type	Organ	Cell type	References
IL-7–YFP	BAC Tg	Thymus	TEC	Alves et al. (2009)
IL-7–ECFP	BAC Tg	Bone marrow	Reticular stromal cell around the sinusoid	Mazzucchelli et al. (2009)
		Thymus	cTEC and mTEC	
IL-7–hCD2	BAC Tg	Fetal thymus	TEC	Repass et al. (2009)
IL-7–Cre	BAC Tg	LN	Unknown	Repass et al. (2009)
		Liver	Unknown	
		Intestine	Unknown	
IL-7–EGFP	Knock-in	Bone marrow	VCAM-1$^+$CD31$^-$ reticular cell	Hara et al. (2012)
		Thymus	cTEC and mTEC	
		LN	gp38$^+$CD31$^+$ LEC, gp38$^+$CD31$^-$ FRC	
		Intestine	IEC	
		Bone marrow	CAR cell	Cordeiro Gomes et al. (2016)
		Fetal liver	VCAM-1$^+$CD105^{low}ALCAM^{high} stromal cell	Tsuneto et al. (2013)
IL-7–EGFP	Knock-in	LN	gp38$^+$CD31$^+$ LEC	Miller et al. (2013)
IL-15–EmGFP	BAC Tg	Spleen	Macrophage, DC, Monocyte, Granulocyte	Colpitts et al. (2012, 2013)
			Mast cell, NK cell, CD8 T cell	
		Bone marrow	HSC	
		Thymus	ETP	
IL-15–EGFP	BAC Tg	Spleen, LN	CD8$^+$ DC	Sosinowski et al. (2013)
IL-15–ECFP	Knock-in	Bone marrow	VCAM-1$^+$PDGFRβ^+CD31$^-$ reticular cell	Cui et al. (2014)

(continued)

Table 1 (continued)

Cytokine reporter mice are summarized with the cytokine-producing cell types

Reporter line	Mouse type	Organ	Cell type	References
		Thymus	MHC-II^{high} mTEC > MHC-II^{low} mTEC > cTEC	
		LN	gp38^{-}CD31^{+} BEC, gp38^{+}CD31^{-} FRC	
		Intestine	IEC, Macrophage, BEC	
		LN	CD8^{+} DC, LPS-stimulated macrophage	

BAC, bacterial artificial chromosome; BEC, blood vascular endothelial cell; CAR, CXCL12-abundant reticular; cTEC, cortical thymic epithelial cell; DC, dendritic cell; ETP, early thymic progenitor; FRC, fibroblastic reticular cell; HSC, hematopoietic stem cell; IEC, intestinal epithelial cell; LEC, lymphatic endothelial cell; LN, lymph node; mTEC, medullary thymic epithelial cell signal of IL-7–ECFP BAC transgenic reporter mice was found in reticular stromal cells in the bone marrow, which often reside near blood vessels (Mazzucchelli et al. 2009). That study also showed that cortical (c) and medullary (m)TECs produce IL-7, but no signal was found in peripheral lymphoid organs. Lastly, IL-7–hCD25 BAC transgenic reporter mice expressed a signal in the fetal thymus (Repass et al. 2009). IL-7/hCD2 mRNA was also detected in the lymph nodes and liver. The same study established IL-7–Cre BAC transgenic mice and crossed them with Rosa26LSL-EYFP mice. EYFP-lineage trace signals were detected in the thymus, lymph nodes, liver, and small intestine. These observations suggest that BAC transgenes may lack the regulatory DNA sequences necessary to express IL-7 in peripheral lymphoid organs, suggesting two types of regulatory sequences: one for IL-7 expression in primary lymphoid organs and the other for IL-7 expression in secondary lymphoid organs.

To distinguish these sequences, we reported an IL-7–GFP knock-in mouse line in which EGFP cDNA was inserted downstream of the translation initiation codon of the endogenous IL-7 locus (Hara et al. 2012). The IL-7/GFP signal was detected in TECs and VCAM-1^+ bone marrow mesenchymal stromal cells, but cTECs expressed higher levels of IL-7 than mTECs (Fig. 1a). The IL-7/GFP signal was also detected in ER-$TR7^+$ mesenchymal stromal cells in the thymus. About two-thirds of VCAM-1^+CD31^- bone marrow reticular stromal cells expressed IL-7/GFP in IL-7–GFP heterozygote mice. Interestingly, the IL-7–GFP knock-in mice also expressed IL-7/GFP in the peripheral lymphoid organs and other tissues including $gp38^+CD31^-$ FRCs of the lymph nodes, $gp38^+CD31^+$ lymphatic endothelial cells (LECs), and IECs (Fig. 1b, c). IL-7 expression in LECs had never been reported previously. The Melchers group later reported an IL-7/GFP signal in VCAM-$1^+CD105^{low}ALCAM^{high}$ stromal cells in the fetal liver of these mice (Tsuneto et al. 2013). In another IL-7–GFP knock-in mouse line, the McCune group reported that LECs express IL-7/GFP at higher levels than FRCs (Miller et al. 2013). They also suggested that IL-7 concentration in lymph was higher than in blood based on phosphorylated STAT5 levels in T cells. These results indicate that IL-7–GFP knock-in mice reflect the expression of endogenous IL-7 more accurately than IL-7 BAC transgenic reporter mice.

Similar to IL-7, to monitor the distribution of IL-15-producing stromal cells in tissues, IL-15 reporter mice have been established (Table 1). The first IL-15–EmGFP BAC transgenic mice exhibited higher and lower levels of IL-15/EmGFP expression in $CD8^+$ DCs and $CD8^-$ DCs, respectively (Colpitts et al. 2012). Vesicular stomatitis virus infection of the reporter mice enhanced the IL-15/EmGFP expression in an IFN-α-dependent manner. Furthermore, monocytes, neutrophils, basophils, eosinophils, and mast cells expressed IL-15/EmGFP at higher levels, while NK cells, NKT cells, $CD8^+$ T cells, marginal zone B cells, and plasmacytoid DCs expressed IL-15/EmGFP at lower levels (Colpitts et al. 2013). In addition, $Lin^-Sca\text{-}1^+c\text{-}kit^+$ hematopoietic stem cells (HSCs), early thymic progenitors (ETPs), and DN2 and DN3 stage thymocytes expressed IL-15/EmGFP. Other IL-15 translational reporter mice, in which an IL-15 BAC transgene has the EGFP reporter joined to the IL-15 coding region via a 2A peptide sequence (IL-15TE), revealed IL-15/EGFP expression in $CD8^+$ DCs, Langerhans cells, and skin-derived $CD103^+$ DCs (Sosinowski et al. 2013).

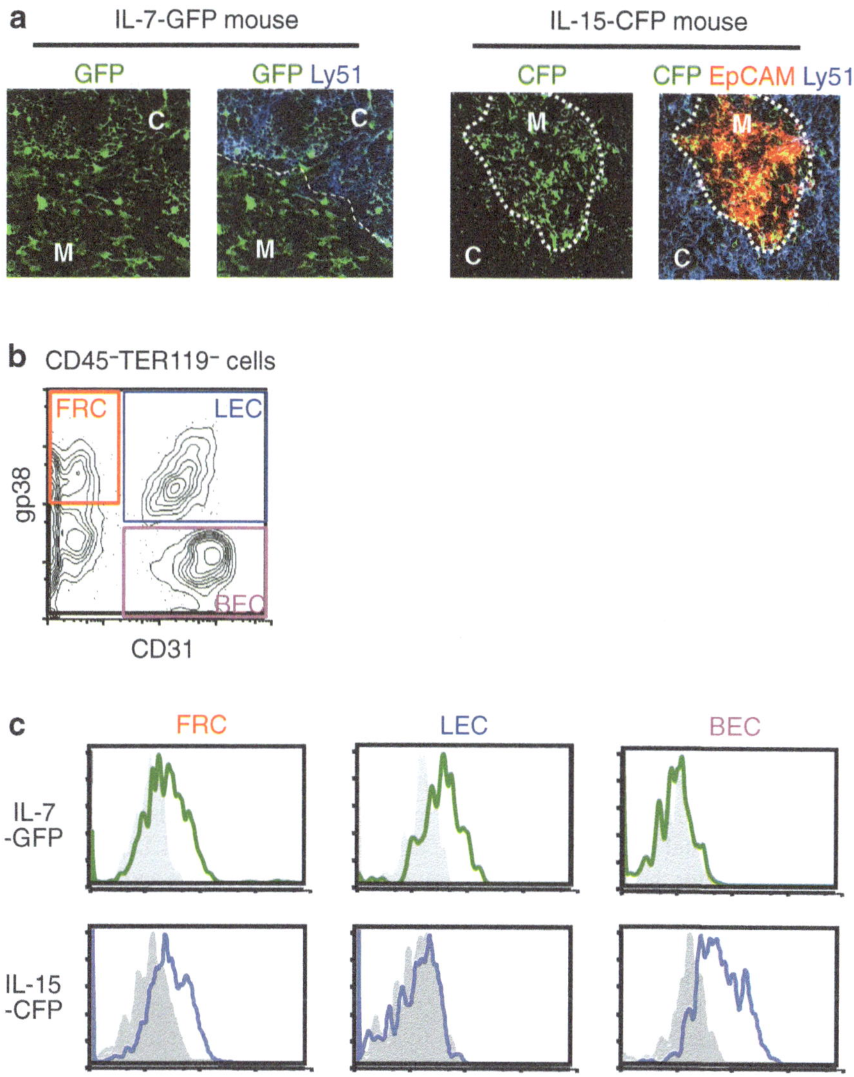

Fig. 1 Identification of stromal cells producing IL-7 and IL-15 by reporter mice. a. Immunohistochemistry of the thymus of IL-7-GFP and IL-15-CFP reporter mice. C, cortex; M, medulla. **b**. Flow cytometric analysis of $CD45^{-}TER119^{-}$ cells in the lymph nodes of IL-7 and IL-15 reporter mice. BEC, blood endothelial cell; FRC, fibroblastic reticular cell; LEC, lymphatic endothelial cell. **c**. Flow cytometric analysis of IL-7/GFP and IL-15/CFP signals detected with anti-GFP antibody in FRC, LEC, and BEC in the lymph nodes of IL-7 and IL-15 reporter mice. Reproduced from (Hara et al. 2012 and Cui et al. 2014)

We established an IL-15–CFP knock-in reporter mouse line in which ECFP cDNA was inserted downstream of the translation initiation codon of the endogenous IL-15 locus (Cui et al. 2014). These mice expressed the IL-15/CFP signal in subsets of mTECs including MHC class II^{hi} mTECs (Fig. 1a), suggesting that the development of IL-15-dependent thymocytes may depend on MHC class II^{hi} mTECs. In the bone marrow, the IL-15/CFP signal was detected in VCAM-1^+PDGFRβ^+CD31^- bone marrow mesenchymal stromal cells. Using IL-7–GFP and IL-15–CFP double reporter mice, we showed many IL-15/CFP$^+$ stromal cells overlapped with IL-7/GFP$^+$ stromal cells. In lymph nodes, fibroblastic reticular cells (FRCs) and high endothelial venules (HEVs) highly expressed IL-15/CFP. Thus, we concluded that several mesenchymal stromal cell types express both IL-7 and IL-15. One of the unique features of IL-15 is its expression in hematopoietic cells such as monocytes, macrophages, and DCs. In our IL-15–CFP mice, IL-15/CFP was detected in activated but not resting macrophages and CD8$^+$ DCs. In the intestines, IL-15/CFP expression was detected in IECs and BECs in the lamina propria, and in lymph nodes, IL-15/CFP was expressed in BECs but not in LECs (Fig. 1b, c). This is in sharp contrast to our IL-7–GFP knock-in mice mentioned above, in which IL-7 was expressed in LECs but not in BECs (Hara et al. 2012). Thus, IL-7 and IL-15 are expressed overlappingly and differentially in lymphoid organs (Cui et al. 2014).

4 Bone Marrow Niche for Early B Cell Development

B cell development takes place in the bone marrow of mice or humans. The early stages of B cell development depend on mesenchymal stromal cells producing the cytokines essential for early B cells. SCF (c-kit ligand) and CXCL12 are two major factors crucial for the expansion of common lymphoid progenitors and pro-B cells. At later stages, IL-7 plays an important role in the expansion of pro-B and early pre-B cells. In 1982, Whitlock and Witte reported that pre-B cells are differentiated from early precursors in a long-term bone marrow culture known as Whitlock–Witte culture (Whitlock and Witte 1982). Later, stromal cell lines supporting B cell development were established from this culture (Whitlock et al. 1987). From the culture supernatant of a stromal cell line, IL-7 was identified as supporting the proliferation of pre-B cells (Namen et al. 1988). Later, the IL-7R gene was identified, and IL-7R was found to be expressed in early B and T cells and mature T cells (Sudo et al. 1993). Isolation of an anti-IL-7R blocking antibody demonstrated that IL-7 is essential for early B and T cell development in the bone marrow and thymus, respectively (Sudo et al. 1993).

The question of the niche for early B cells in the bone marrow went unsolved for some time. Tokoyoda et al. reported that pre-pro-B cells contact CXCL12-abundant reticular (CAR) cells, whereas pro B cells move away from these cells to interact with other IL-7-expressing stromal cells (Tokoyoda et al. 2004). They used an anti-IL-7 antibody to detect the IL-7 expression in stromal cells. In another study, osteoblast-specific Col2.3-Cre CXCL12 cKO mice exhibited about a 50% reduction of common lymphoid progenitors (CLPs) in the bone marrow, whereas the number of pre-pro-B, pro-B, and pre-B cells was unchanged (Ding and Morrison 2013). By contrast, endothelial cell- and hematopoietic cell-specific Tie2-Cre and CAR cell-specific Lepr-Cre CXCL12 cKO mice showed unchanged levels of CLPs, pre-pro-B, pro-B, and pre-B cells, but Prx1-Cre CXCL12 cKO mice, which lack CXCL12 in all types of mesenchymal cells, showed severely reduced numbers in the bone marrow. In addition, $Lin^{-}IL\text{-}7R^{+}$ cells, which include CLPs, were enriched at the endosteum of wild-type mice. These results suggested that osteoblasts are the niche for CLPs. A recent study reported that Lepr-Cre but not Tie2-Cre SCF cKO mice exhibited a significant reduction in CLPs in the bone marrow and that Lepr-Cre SCF cKO mice showed no change in pro-B or pre-B cell numbers (Comazzetto et al. 2019). However, HSCs were also reduced in the bone marrow of both mouse models, complicating the conclusions (Ding et al. 2012). Thus, which cells, osteoblasts or CAR cells, are essential for CLPs remains controversial.

Lepr-Cre IL-7 cKO mice exhibited a severe reduction of B cell lineage-committed $Ly6D^{+}$ CLPs, pro-B cells, and pre-B cells in the bone marrow, and most of their $IL\text{-}7^{+}$ stromal cells were CAR cells, suggesting that CAR cells play an important role in early B cell development (Cordeiro Gomes et al. 2016) (Table 2, Fig. 2). In addition, Tie2-Cre IL-7 cKO mice exhibited a mild decrease in pro-B and pre-B cells, suggesting that BEC-derived IL-7 has a minor role in supporting B cell development in the bone marrow. Consistently, Fistonich et al. showed that $Rag2\text{-}GFP^{high}$ pro-B cells directly contacted $IL\text{-}7^{+}CXCL12^{+}$ CAR cells more than $Rag2\text{-}GFP^{low}$ pre-B cells (Fistonich et al. 2018). Furthermore, Balzano et al. recently reported that TdT^{+} pro-B cells closely contacted $IL\text{-}7^{+}$ perivascular stromal cells but not osteoblasts in the endosteum (Balzano et al. 2019). These results suggest that CAR cells in the parenchymal and perivascular regions might act as the niche for CLPs and early B precursors in the bone marrow.

Adiponectin is a hormone produced by adipocytes, which elevates insulin sensitivity of the liver and muscle. In addition, adiponectin plays roles in the immune system. Indeed, it inhibited the proliferation of myeloid cells and early B cells, suppressed macrophages and NK cells, and increased HSC proliferation (DiMascio et al. 2007; Kim et al. 2006; Yokota et al. 2000). Previously, it was suggested that adiponectin was expressed in $Lepr^{+}$ stromal cells in the bone marrow (Zhou et al. 2017) and later confirmed that it was expressed by CAR cells (Mukohira et al. 2019). Additionally, adiponectin (Adipoq)-Cre IL-7 cKO mice exhibited a severe reduction of pro-B and pre-B cells but not pre-pro-B cells in the bone marrow (Mukohira et al. 2019), suggesting that CAR cell-derived IL-7 is essential for early B cell development (Table 2). This study confirmed the notion that IL-7-producing CAR cells are the niche for early B cells in the bone marrow.

Table 2 Local function of IL-7 and IL-15 in cell-type-specific cKO mice

Mouse line	Cre driver	Targeted cell	Target cell of cytokine action	Function	References
IL-7-floxed	FoxN1-Cre	TEC	Thymocyte, IEL	Survival, Proliferation	Shitara et al. (2013)
	Alb-Cre	Hepatocyte	T cell, NKT cell	Survival	Liang et al. (2012)
Lepr-Cre	CAR cell	$Ly6D^+$ CLP, Pro-B cell, Pre-B cell	Survival, Proliferation	Cordeiro	Gomes et al. (2016)
	Adipoq-Cre	CAR cell	Pro-B cell, Pre-B cell	Survival, Proliferation	Mukohira et al. (2019)
	K5-Cre	keratinocyte	Memory $CD4^+$ and $CD8^+$ T cell	Maintenance	Adachi et al. (2015)
	Tie2-Cre	LEC	Memory $CD4^+$ Th2 cell	Maintenance	Shinoda et al. (2016)
IL-15Rα-floxed	LysM-Cre	Macrophage	NK cell, Memory $CD8^+$ T cell	Proliferation, Maturation	Mortier et al. (2009)
	CD11c-Cre	DC	NK cell, Memory $CD8^+$ T cell	Proliferation, Maturation	ibid.
	Vil-Cre	IEC	$CD8\alpha\alpha^+$ IEL	Maintenance	ibid.
IL-15-floxed	Adipoq-Cre	Adipocyte	NK cell	Survival	Liou et al. (2014)
	LysM-Cre	Macrophage	NK cell	Survival	ibid.
	CD11c-Cre	DC	NK cell	Survival	ibid.
	Vil-Cre	IEC	$CD8\alpha\alpha^+$ IEL, NK cell in LP	Survival, Proliferation	Zhu et al. (2020)
	Ccl19-Cre	FRC	NK cell and ILC1	Maintenance	Gil-Cruz et al. (2016)

Alb, albumin; Adipoq, adiponectin; CAR, CXCL12-abundant reticular; CLP, common lymphoid progenitor; DC, dendritic cell; FRC, fibroblastic reticular cell; IEC, intestinal epithelial cell; IEL, intraepithelial lymphocyte; ILC1, group 1 innate lymphoid cell; K5, keratin 5; LEC, lymphatic endothelial cell; Lepr, leptin receptor; LP, lamina propria; TEC, thymic epithelial cell; Vil, villin

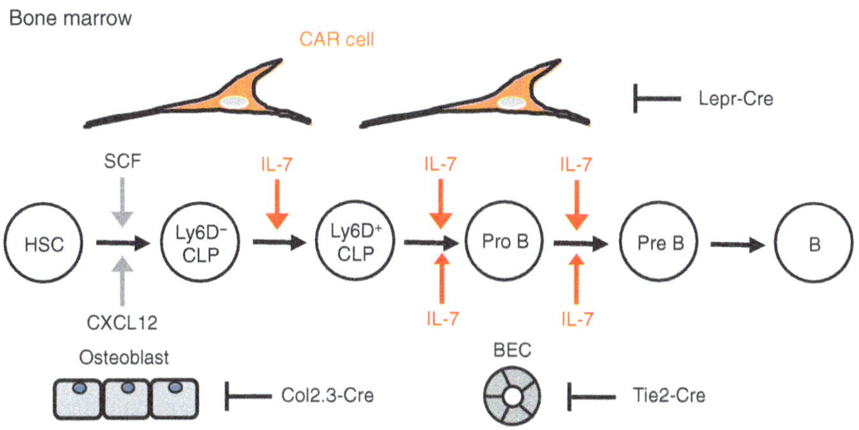

Fig. 2 Bone marrow niche for early B cell development. A schema depicting the bone marrow niche for early B cell development. The three types of Cre-driver mouse lines are associated with different target stromal cell types. Vertical arrows indicate the possible involvement of cytokines in each step of B cell differentiation. CAR cells and BECs play major and minor roles, respectively, in IL-7-dependent stages of early B cell development. BEC, blood vascular endothelial cell; CAR, CXCL12-abundant reticular; CLP, common lymphoid progenitor; HSC, hematopoietic stem cell

5 Local Roles of IL-7 in Lymphoid and Non-Lymphoid Organs

To address the function of IL-7 in the local microenvironment, an IL-7-floxed mouse line was established and crossed with various Cre-driver mice (Table 2). IL-7 is produced by TECs and mesenchymal stromal cells in the thymus. To clarify the role of TEC-derived IL-7, thymocyte development was examined in FoxN1-Cre IL-7 cKO mice (Shitara et al. 2013). The development was severely impaired, but the reduction of thymocytes was not as severe as in IL-7 KO mice, suggesting that TEC-derived IL-7 plays a major role in thymocyte development and that mesenchymal stromal cell-derived IL-7 plays a minor role. Interestingly, TCR$\gamma\delta^+$ IELs of the small intestine were also severely reduced in number in FoxN1-Cre IL-7 cKO mice. This result supports the idea that $\gamma\delta$ IELs originate from IEL precursors in the thymus, suggesting the thymic origin of $\gamma\delta$ IELs.

The liver contains many immune cells such as T cells, NKT cells, NK cells, and macrophages (Racanelli and Rehermann 2006). It is reported that hepatocytes produce IL-7 (Sawa et al. 2009), although no IL-7/GFP signal was detected in IL-7–GFP reporter mice (Hara et al. 2012). It is also reported that hepatocyte-derived IL-7 plays a role in inflammation (Sawa et al. 2009). On the other hand, intrahepatic iNKT cells mainly depend on IL-15 for their maintenance (Matsuda et al. 2002). In Alb-Cre IL-7 cKO mice, both T and NKT cell numbers were reduced in the liver, but NK cells were unchanged (Liang et al. 2012). Because NKT cells reside within the sinusoidal cavity in the liver, they might receive hepatocyte-derived IL-7 that is bound to endothelial cells of the sinusoid or through cell extrusions of hepatocytes. Thus, this study revealed a novel function of hepatocyte-derived IL-7 in the maintenance of NKT cells, implying that iNKT cells might depend on both IL-7 and IL-15 in the liver.

Memory $CD8^+$ T cells depend on IL-7 and IL-15 and are maintained in the bone marrow in addition to peripheral lymphoid organs. Sercan Alp et al. addressed the question of whether memory-phenotype $CD8^+$ T cells interact with $IL\text{-}7^+$ stromal cells in the bone marrow (Sercan Alp et al. 2015). They found that the majority of $CD44^+CD8^+$ T cells were either in direct contact with or within 10 μm of $IL\text{-}7/GFP^+$ stromal cells. This result indicates that the IL-7-producing stromal cells might be the niche for memory $CD8^+$ T cells in the bone marrow of mice. However, future studies with IL-7 cKO mice are necessary to test this hypothesis.

The skin harbors resident memory T cells important for long-term skin immunity. In the skin, hair follicle keratinocytes expressed IL-7 and IL-15, and memory $CD4^+$ and $CD8^+$ T cells localized near hair follicles (Adachi et al. 2015). The study also showed that the keratinocyte-specific deletion of IL-7 in keratin 5 (K5)-Cre IL-7 cKO mice caused impaired maintenance of memory $CD4^+$ and $CD8^+$ T cells. Further experiments in that work using cell transfers revealed that IL-15 produced by keratinocytes was indispensable for the repopulation of memory $CD8^+$ T cells in the skin. These results suggested that hair follicle keratinocytes might be the niche for skin-resident memory $CD4^+$ and $CD8^+$ T cells.

Memory CD4 T cells depend on IL-7 for their maintenance. To maintain memory CD4 T cells at local inflammatory sites after chronic allergic inflammation, tertiary lymphoid structures called inducible bronchus-associated lymphoid tissue (iBALT) were generated in the lung (Shinoda et al. 2016). Additionally, pathogenic memory Th2 cells were maintained in the iBALT, and LECs produced a high level of IL-7 in iBALT. The study also showed that in Tie2-Cre IL-7 cKO mice with a specific deletion of IL-7 in endothelial cells, the size of iBALT and the number of memory Th2 cells in iBALT were reduced. These results suggested that IL-7-producing LECs in iBALT might be the niche for memory Th2 cells.

6 Niches for Early NK Cell Development in Bone Marrow

NK cell development takes place in the bone marrow and depends on IL-15 (Ma et al. 2006). As described above, IL-15 is expressed by certain CAR cells, BECs, and myeloid cells in the bone marrow (Cui et al. 2014; Noda et al. 2011), making these cells candidates for the IL-15 niche of early NK cells. Noda et al. reported that CAR cells expressed IL-15 and IL-15Rα (Noda et al. 2011). They also found that Mx-Cre CXCR4 cKO mice exhibited a reduction of NK cells in the bone marrow after polyI:polyC injection, suggesting that CAR cell-derived CXCL12 might be important for NK cells. Mortier et al. reported that $CD27^-CD11b^+$ mature NK cells, but not immature NK cells, were significantly reduced in the bone marrow of LysM-Cre and

CD11c-Cre IL-15Rα cKO mice (Mortier et al. 2009), suggesting that the IL-15/IL-15Rα complex produced by macrophages and DCs is important for early NK cell development in the bone marrow. Castillo et al. took a slightly different approach (Castillo et al. 2009). They established IL-15Rα transgenic mice that express IL-15Rα in DCs (DC/IL-15Rα) and crossed them with IL-15Rα KO mice. In the KO mice, NK cells were drastically reduced in the bone marrow, liver, and spleen. Introduction of the DC/IL-15Rα transgene partially rescued immature and mature NK cells, suggesting a potential role of IL-15Rα expressed by DCs in the maintenance of NK cells. Because some mature NK cells in the bone marrow may be circulating and depend on IL-15/IL-15Rα in peripheral organs, it is not clear whether NK cell development depends on IL-15 produced by myeloid cells in the bone marrow. Thus, functional experiments are needed to conclude the source of IL-15 (CAR cells or myeloid cells) that regulates NK cell development.

7 Local Roles of IL-15 in the Periphery

The local function of IL-15 has been addressed with IL-15Rα-floxed mice (Mortier et al. 2009). $CD27^{-}CD11b^{+}$ mature NK cells were reduced in the bone marrow and spleen of LysM-Cre and CD11c-Cre IL-15Rα cKO mice, whereas the number of $CD27^{+}CD11b^{-}$ and $CD27^{+}CD11b^{+}$ immature NK cells was unchanged. The decrease of $CD27^{-}CD11b^{+}$ mature NK cells was attributed to reduced proliferation of $CD27^{+}CD11b^{+}$ NK cells. Furthermore, memory phenotype $CD8^{+}$ T cells were also reduced in the bone marrow, spleen, and lymph nodes of both mouse models. These results indicate that IL-15 produced by macrophages and DCs is important for the proliferation of NK cells and the maintenance of memory $CD8^{+}$ T cells.

The local function of IL-15 was also addressed with IL-15-floxed mice (Table 2). Adipoq-Cre IL-15 cKO mice with a specific deletion of IL-15 from adipocytes have a significantly reduced number of NK cells in the spleen, liver, and visceral adipose tissue, suggesting that IL-15 produced by adipocytes might control the maintenance of local and systemic NK cells (Liou et al. 2014). The cKO did not affect the number of immature or mature NK cells in the bone marrow, however. Because the Adipoq-Cre transgene also targets CAR cells in the bone marrow (Mukohira et al. 2019), there is the possibility that NK cell development in the bone marrow does not depend on IL-15 produced by CAR cells. Indeed, CD11c-Cre and LysM-Cre IL-15 cKO mice with a specific deletion of IL-15 from DCs and macrophages, respectively, showed fewer systemic NK cells, but not to the degree seen in Adipoq-Cre IL-15 cKO mice (Liou et al. 2014).

The local function of IL-15 in the liver has been addressed in Alb-Cre IL-15Rα cKO mice with a specific deletion of IL-15 in hepatocytes (Cepero-Donates et al. 2016). These mice showed a significant drop in NK and iNKT cells in the liver, suggesting that intrahepatic NK and iNKT cells depend on local IL-15 for their maintenance. The same study also found that NK and NKT cells were reduced in LysM-Cre IL-15Rα cKO mice. These results suggest that IL-15 derived from hepatocytes and macrophages regulates NK and NKT cells. Because hepatocyte-derived IL-7 is important for the maintenance of NKT cells in the liver (Liang et al. 2012), it is possible that intrahepatic NKT cells depend on both IL-7 and IL-15 produced by hepatocytes. Additionally, because the NK cell fraction which they analyzed contained both conventional NK cells and ILC1s (Sojka et al. 2014), further studies are needed to clarify the IL-15 niches for these cell populations.

As for the local function of IL-15 in the intestines, IEC-specific Vil-Cre IL-15Rα cKO mice exhibited a significant reduction of CD8α^{+} γδ IELs, suggesting that IL-15Rα in IECs is essential for the maintenance of CD8αα IELs (Mortier et al. 2009). Vil-Cre IL-15 cKO mice with a specific deletion of IL-15 from IECs had fewer CD8αα^{+} γδ and αβ IELs, but the number of CD8αβ^{+} αβ IELs was unaffected (Zhu et al. 2020). By contrast, the IEL count was unchanged in Tie2-Cre and LysM-Cre IL-15 cKO mice with specific deletion of IL-15 in BECs and macrophages, respectively, implying that IL-15 produced by these cell types does not play a role in the maintenance of IELs. NK cells in the lamina propria were also reduced in Vil-Cre IL-15 cKO mice, suggesting that IEC-derived IL-15 might play a role in the maintenance of NK cells in the lamina propria of the intestines.

IL-15 is produced by various stromal cells in the lymph nodes. Ccl19-Cre IL-15 cKO mice with the specific deletion of IL-15 in FRCs of the lymph nodes have no CD3^{-}NK1.1^{+}NKp46^{+} NK cells or ILC1s in the lymph nodes and Peyer's patches (Gil-Cruz et al. 2016). After infection with a mouse hepatitis virus, Ccl19-Cre MyD88 cKO mice with the deletion of innate sensing adaptor MyD88 in FRCs exhibited a higher IL-15 expression in stromal cells in Peyer's patches than control mice. Furthermore, the increased IL-15 induced the expansion of NK cells and ILC1s in the Ccl19-Cre MyD88 cKO mice, resulting in an accelerated decrease of viral titers. Thus, these results suggested that FRCs might regulate the population size of NK cells and ILC1s via IL-15, thereby balancing infection control and tissue repair.

8 Cytokine Expression in Stromal Cells Revealed by ScRNA-Seq Analysis

ScRNA-seq has facilitated to dissect different subpopulations of stromal cells in the lymphoid organs. The Aifantis group sorted three populations of bone marrow stromal cells, VE-Cad-tdTomato^{+} endothelial cells, Lepr-tdTomato^{+} mesenchymal stromal cells, and Col2.3-tdTomato^{+} osteoblasts, and carried out scRNA-seq on each population (Tikhonova et al. 2019). Lepr^{+} mesenchymal stromal cells were further divided into four clusters, P1, P2, P3, and P4. *Mgp*^{high} P1 and *Lpl*^{high} P2 clusters expressed adipogenesis-associated markers and covered sinusoidal capillaries, while *Wif1*^{high} P3 and *Spp1*^{high}*Ibsp*^{high} P4 clusters expressed osteogenesis-associated markers and were localized to the trabecular portion of the bone. The P1, P2, P3, and P4 clusters represented 55%, 14%, 13%, and 18%, respectively, of Lepr^{+} mesenchymal stromal cells. Interestingly, IL-7 mRNA was highly expressed in P2 and moderately in P1 and P3 clusters, suggesting that 82% of $Lepr^+$ mesenchymal stromal cells might express IL-7 mRNA. This result is similar to previous findings that about 60% of VCAM-1^+PDGFRβ^+ mesenchymal stromal cells express IL-7 in IL-7–GFP reporter mice (Cordeiro Gomes et al. 2016; Hara et al. 2012; Mukohira et al. 2019). The scRNA-seq analysis further showed that IL-15 mRNA was expressed mainly in the P1 and P2 clusters (Tikhonova et al. 2019). Importantly, expressions of CXCL12 and SCF, which are important for the maintenance of HSCs and hematopoietic progenitors, were mainly detected in the P1, P2, and P3 clusters, which is also consistent with previous reports (Ding and Morrison 2013; Omatsu et al. 2010).

Lymph node stromal cells have been divided into four groups based on podoplanin (PDPN, or gp38) and CD31 expression: $PDPN^+CD31^-$ FRCs, $PDPN^-CD31^-$ double-negative cells (DNCs), $PDPN^+CD31^+$ LECs, and $PDPN^-CD31^+$ BECs (Malhotra et al. 2012). To report the gene expression profiles of single stromal cells in lymph nodes, one study sorted $CD45^-CD31^-$ non-endothelial stroma cells, which include FRCs and DNCs and carried out scRNA-seq analysis (Rodda et al. 2018). The study divided the stromal cell into 9 clusters: follicular dendritic cells (FDCs), perivascular cells (PvCs), marginal reticular cells (MRCs), $CD34^+$ stromal cells, $Cxcl9^+$ T-zone reticular cells (TRCs), $Ccl9^{lo}$ TRCs, $Ccl9^{hi}$ TRCs, $Inmt^+$ stromal cells, and $Nr4a1^+$ stromal cells. Among them, IL-7 was expressed in three TRC groups, $Nr4a1^+$ stromal cells, $Inmt^+$ stromal cells, and MRCs. This result is consistent with our result that a part of FRCs in the lymph nodes expresses IL-7 in IL-7–GFP reporter mice (Hara et al. 2012).

Although scRNA-seq provides a new approach to study the heterogeneity of different cell subsets, it still has several limitations such as high technical noise, low coverage, and amplification bias. Furthermore, scRNA-seq data have no spatial information. To overcome these limitations, Baccin et al. recently reported a new comprehensive method that combines scRNA-seq and laser-capture microdissection coupled with sequencing (LCM-seq) (Baccin et al. 2020). They could thus divide CAR cells into two subsets, Adipo-CAR and Osteo-CAR cells that differentially localize to sinusoidal and arteriolar surfaces, respectively. The incorporation of spatial transcriptome analysis will advance niche research.

9 Conclusion

In this review, we describe the current understanding of several cytokine-producing immune microenvironments in lymphoid organs. Despite remarkable achievements, there still remain many questions. For example, it is still unclear which stromal cells or hematopoietic cells are the niches for the development of early NK cells and ILCs in the bone marrow and fetal liver. Because NK cells and ILCs depend on IL-15 or IL-7 for their differentiation, cell-type-specific gene knockout mice are recommended. Moreover, stromal cells in peripheral non-lymphoid tissues deserve more attention. Lymphocytes that reside in peripheral tissues include memory T and

B cells, some γδ T cells, iNKT cells, and ILCs, which all play important roles in innate and adaptive immunity and immune diseases. It is largely unknown which stromal cells are involved in the maintenance and function of these tissue-resident lymphocytes in the mucosa and skin. Because some tissue-resident lymphocytes also depend on IL-7 or IL-15, cell-type-specific gene knockout mice will be effective for these studies too.

Acknowledgements This work was supported by the Japan Society for the Promotion of Science (JSPS) KAKENHI Grants 16H05172 and 20H03501 (K.I.), 16K08835 and 19K07604 (T.H.), and 19K16687 and 17K15721 (G.C.); a grant from the Takeda Science Foundation (G.C.); and from the Shimizu Foundation for Immunology and Neuroscience grant for 2016 (G.C.). It was also supported by the Joint Usage Research Center program of the Institute for Frontier Life and Medical Sciences, Kyoto University.

References

Adachi S, Yoshida H, Honda K, Maki K, Saijo K, Ikuta K, Saito T, Nishikawa SI (1998) Essential role of IL-7 receptor α in the formation of Peyer's patch anlage. Int Immunol 10:1–6

Adachi T, Kobayashi T, Sugihara E, Yamada T, Ikuta K, Pittaluga S, Saya H, Amagai M, Nagao K (2015) Hair follicle-derived IL-7 and IL-15 mediate skin-resident memory T cell homeostasis and lymphoma. Nat Med 21:1272–1279

Alves NL, Richard-Le Goff O, Huntington ND, Sousa AP, Ribeiro VS, Bordack A, Vives FL, Peduto L, Chidgey A, Cumano A, Boyd R, Eberl G, Di Santo JP (2009) Characterization of the thymic IL-7 niche in vivo. Proc Natl Acad Sci USA 106:1512–1517

Baccin C, Al-Sabah J, Velten L, Helbling PM, Grunschlager F, Hernandez-Malmierca P, Nombela-Arrieta C, Steinmetz LM, Trumpp A, Haas S (2020) Combined single-cell and spatial transcriptomics reveal the molecular, cellular and spatial bone marrow niche organization. Nat Cell Biol 22:38–48

Balzano M, De Grandis M, Vu Manh TP, Chasson L, Bardin F, Farina A, Serge A, Bidaut G, Charbord P, Herault L, Bailly AL, Cartier-Michaud A, Boned A, Dalod M, Duprez E, Genever P, Coles M, Bajenoff M, Xerri L, Aurrand-Lions M, Schiff C, Mancini SJC (2019) Nidogen-1 contributes to the interaction network involved in pro-B cell retention in the peri-sinusoidal hematopoietic stem cell niche. Cell Rep 26:3257–3271 e3258

Castillo EF, Stonier SW, Frasca L, Schluns KS (2009) Dendritic cells support the in vivo development and maintenance of NK cells via IL-15 trans-presentation. J Immunol 183:4948–4956

Cepero-Donates Y, Rakotoarivelo V, Mayhue M, Ma A, Chen YG, Ramanathan S (2016) Homeostasis of IL-15 dependent lymphocyte subsets in the liver. Cytokine 82:95–101

Colpitts SL, Stoklasek TA, Plumlee CR, Obar JJ, Guo C, Lefrancois L (2012) The role of IFN-α receptor and MyD88 signaling in induction of IL-15 expression in vivo. J Immunol 188:2483–2487

Colpitts SL, Stonier SW, Stoklasek TA, Root SH, Aguila HL, Schluns KS, Lefrancois L (2013) Transcriptional regulation of IL-15 expression during hematopoiesis. J Immunol 191:3017–3024

Comazzetto S, Murphy MM, Berto S, Jeffery E, Zhao Z, Morrison SJ (2019) Restricted hematopoietic progenitors and erythropoiesis require SCF from leptin receptor+ niche cells in the bone marrow. Cell Stem Cell 24:477–486 e476

Cordeiro Gomes A, Hara T, Lim VY, Herndler-Brandstetter D, Nevius E, Sugiyama T, Tani-ichi S, Schlenner S, Richie E, Rodewald HR, Flavell RA, Nagasawa T, Ikuta K, Pereira JP (2016) Hematopoietic stem cell niches produce lineage-instructive signals to control multipotent progenitor differentiation. Immunity 45:1219–1231

Cui G, Hara T, Simmons S, Wagatsuma K, Abe A, Miyachi H, Kitano S, Ishii M, Tani-ichi S, Ikuta K (2014) Characterization of the IL-15 niche in primary and secondary lymphoid organs in vivo. Proc Natl Acad Sci USA 111:1915–1920

Diefenbach A, Colonna M, Koyasu S (2014) Development, differentiation, and diversity of innate lymphoid cells. Immunity 41:354–365

DiMascio L, Voermans C, Uqoezwa M, Duncan A, Lu D, Wu J, Sankar U, Reya T (2007) Identification of adiponectin as a novel hemopoietic stem cell growth factor. J Immunol 178:3511–3520

Ding L, Morrison SJ (2013) Haematopoietic stem cells and early lymphoid progenitors occupy distinct bone marrow niches. Nature 495:231–235

Ding L, Saunders TL, Enikolopov G, Morrison SJ (2012) Endothelial and perivascular cells maintain haematopoietic stem cells. Nature 481:457–462

Fistonich C, Zehentmeier S, Bednarski JJ, Miao R, Schjerven H, Sleckman BP, Pereira JP (2018) Cell circuits between B cell progenitors and IL-7^{+} mesenchymal progenitor cells control B cell development. J Exp Med 215:2586–2599

Gil-Cruz C, Perez-Shibayama C, Onder L, Chai Q, Cupovic J, Cheng HW, Novkovic M, Lang PA, Geuking MB, McCoy KD, Abe S, Cui G, Ikuta K, Scandella E, Ludewig B (2016) Fibroblastic reticular cells regulate intestinal inflammation via IL-15-mediated control of group 1 ILCs. Nat Immunol 17:1388–1396

Grabstein KH, Eisenman J, Shanebeck K, Rauch C, Srinivasan S, Fung V, Beers C, Richardson J, Schoenborn MA, Ahdieh M, Johnson L, Alderson MR, Watson JD, Anderson DM, Giri JG (1994) Cloning of a T cell growth factor that interacts with the β chain of the interleukin-2 receptor. Science 264:965–968

Hara T, Shitara S, Imai K, Miyachi H, Kitano S, Yao H, Tani-ichi S, Ikuta K (2012) Identification of IL-7-producing cells in primary and secondary lymphoid organs using IL-7-GFP knock-in mice. J Immunol 189:1577–1584

Kang J, Coles M (2012) IL-7: the global builder of the innate lymphoid network and beyond, one niche at a time. Semin Immunol 24:190–197

Kim KY, Kim JK, Han SH, Lim JS, Kim KI, Cho DH, Lee MS, Lee JH, Yoon DY, Yoon SR, Chung JW, Choi I, Kim E, Yang Y (2006) Adiponectin is a negative regulator of NK cell cytotoxicity. J Immunol 176:5958–5964

Liang B, Hara T, Wagatsuma K, Zhang J, Maki K, Miyachi H, Kitano S, Yabe-Nishimura C, Tani-ichi S, Ikuta K (2012) Role of hepatocyte-derived IL-7 in maintenance of intrahepatic NKT cells and T cells and development of B cells in fetal liver. J Immunol 189:4444–4450

Link A, Vogt TK, Favre S, Britschgi MR, Acha-Orbea H, Hinz B, Cyster JG, Luther SA (2007) Fibroblastic reticular cells in lymph nodes regulate the homeostasis of naive T cells. Nat Immunol 8:1255–1265

Liou YH, Wang SW, Chang CL, Huang PL, Hou MS, Lai YG, Lee GA, Jiang ST, Tsai CY, Liao NS (2014) Adipocyte IL-15 regulates local and systemic NK cell development. J Immunol 193:1747–1758

Ma A, Koka R, Burkett P (2006) Diverse functions of IL-2, IL-15, and IL-7 in lymphoid homeostasis. Annu Rev Immunol 24:657–679

Malhotra D, Fletcher AL, Astarita J, Lukacs-Kornek V, Tayalia P, Gonzalez SF, Elpek KG, Chang SK, Knoblich K, Hemler ME, Brenner MB, Carroll MC, Mooney DJ, Turley SJ, Immunological Genome Project C (2012) Transcriptional profiling of stroma from inflamed and resting lymph nodes defines immunological hallmarks. Nat Immunol 13:499–510

Matsuda JL, Gapin L, Sidobre S, Kieper WC, Tan JT, Ceredig R, Surh CD, Kronenberg M (2002) Homeostasis of Vα14*i* NKT cells. Nat Immunol 3:966–974

Matsue H, Bergstresser PR, Takashima A (1993) Keratinocyte-derived IL-7 serves as a growth factor for dendritic epidermal T cells in mice. J Immunol 151:6012–6019

Mattei F, Schiavoni G, Belardelli F, Tough DF (2001) IL-15 is expressed by dendritic cells in response to type I IFN, double-stranded RNA, or lipopolysaccharide and promotes dendritic cell activation. J Immunol 167:1179–1187

Mazzucchelli R, Durum SK (2007) Interleukin-7 receptor expression: intelligent design. Nat Rev Immunol 7:144–154
Mazzucchelli RI, Warming S, Lawrence SM, Ishii M, Abshari M, Washington AV, Feigenbaum L, Warner AC, Sims DJ, Li WQ, Hixon JA, Gray DH, Rich BE, Morrow M, Anver MR, Cherry J, Naf D, Sternberg LR, McVicar DW, Farr AG, Germain RN, Rogers K, Jenkins NA, Copeland NG, Durum SK (2009) Visualization and identification of IL-7 producing cells in reporter mice. PLoS One 4:e7637
Miller CN, Hartigan-O'Connor DJ, Lee MS, Laidlaw G, Cornelissen IP, Matloubian M, Coughlin SR, McDonald DM, McCune JM (2013) IL-7 production in murine lymphatic endothelial cells and induction in the setting of peripheral lymphopenia. Int Immunol 25:471–483
Mortier E, Advincula R, Kim L, Chmura S, Barrera J, Reizis B, Malynn BA, Ma A (2009) Macrophage- and dendritic-cell-derived interleukin-15 receptor alpha supports homeostasis of distinct $CD8^+$ T cell subsets. Immunity 31:811–822
Mukohira H, Hara T, Abe S, Tani-ichi S, Sehara-Fujisawa A, Nagasawa T, Tobe K, Ikuta K (2019) Mesenchymal stromal cells in bone marrow express adiponectin and are efficiently targeted by an adiponectin promoter-driven Cre transgene. Int Immunol 31:729–742
Namen AE, Lupton S, Hjerrild K, Wignall J, Mochizuki DY, Schmierer A, Mosley B, March CJ, Urdal D, Gillis S (1988) Stimulation of B-cell progenitors by cloned murine interleukin-7. Nature 333:571–573
Noda M, Omatsu Y, Sugiyama T, Oishi S, Fujii N, Nagasawa T (2011) CXCL12-CXCR4 chemokine signaling is essential for NK-cell development in adult mice. Blood 117:451–458
Omatsu Y, Sugiyama T, Kohara H, Kondoh G, Fujii N, Kohno K, Nagasawa T (2010) The essential functions of adipo-osteogenic progenitors as the hematopoietic stem and progenitor cell niche. Immunity 33:387–399
Racanelli V, Rehermann B (2006) The liver as an immunological organ. Hepatology 43:S54-62
Raeber ME, Zurbuchen Y, Impellizzieri D, Boyman O (2018) The role of cytokines in T-cell memory in health and disease. Immunol Rev 283:176–193
Repass JF, Laurent MN, Carter C, Reizis B, Bedford MT, Cardenas K, Narang P, Coles M, Richie ER (2009) IL7-hCD25 and IL7-Cre BAC transgenic mouse lines: new tools for analysis of IL-7 expressing cells. Genesis 47:281–287
Robinette ML, Bando JK, Song W, Ulland TK, Gilfillan S, Colonna M (2017) IL-15 sustains IL-7R-independent ILC2 and ILC3 development. Nat Commun 8:14601
Rodda LB, Lu E, Bennett ML, Sokol CL, Wang X, Luther SA, Barres BA, Luster AD, Ye CJ, Cyster JG (2018) Single-cell RNA sequencing of lymph node stromal cells reveals niche-associated heterogeneity. Immunity 48:1014-1028.e6
Sawa Y, Arima Y, Ogura H, Kitabayashi C, Jiang JJ, Fukushima T, Kamimura D, Hirano T, Murakami M (2009) Hepatic interleukin-7 expression regulates T cell responses. Immunity 30:447–457
Sercan Alp O, Durlanik S, Schulz D, McGrath M, Grun JR, Bardua M, Ikuta K, Sgouroudis E, Riedel R, Zehentmeier S, Hauser AE, Tsuneto M, Melchers F, Tokoyoda K, Chang HD, Thiel A, Radbruch A (2015) Memory $CD8^+$ T cells colocalize with IL-7^+ stromal cells in bone marrow and rest in terms of proliferation and transcription. Eur J Immunol 45:975–987
Shinoda K, Hirahara K, Iinuma T, Ichikawa T, Suzuki AS, Sugaya K, Tumes DJ, Yamamoto H, Hara T, Tani-ichi S, Ikuta K, Okamoto Y, Nakayama T (2016) $Thy1^+$IL-7^+ lymphatic endothelial cells in iBALT provide a survival niche for memory T-helper cells in allergic airway inflammation. Proc Natl Acad Sci USA 113:E2842-2851
Shitara S, Hara T, Liang B, Wagatsuma K, Zuklys S, Hollander GA, Nakase H, Chiba T, Tani-ichi S, Ikuta K (2013) IL-7 produced by thymic epithelial cells plays a major role in the development of thymocytes and $TCR\gamma\delta^+$ intraepithelial lymphocytes. J Immunol 190:6173–6179
Sojka DK, Plougastel-Douglas B, Yang L, Pak-Wittel MA, Artyomov MN, Ivanova Y, Zhong C, Chase JM, Rothman PB, Yu J, Riley JK, Zhu J, Tian Z, Yokoyama WM (2014) Tissue-resident natural killer (NK) cells are cell lineages distinct from thymic and conventional splenic NK cells. Elife 3:e01659

Sosinowski T, White JT, Cross EW, Haluszczak C, Marrack P, Gapin L, Kedl RM (2013) CD8α^+ dendritic cell trans presentation of IL-15 to naive CD8$^+$ T cells produces antigen-inexperienced T cells in the periphery with memory phenotype and function. J Immunol 190:1936–1947
Sudo T, Ito M, Ogawa Y, Iizuka M, Kodama H, Kunisada T, Hayashi S, Ogawa M, Sakai K, Nishikawa S (1989) Interleukin 7 production and function in stromal cell-dependent B cell development. J Exp Med 170:333–338
Sudo T, Nishikawa S, Ohno N, Akiyama N, Tamakoshi M, Yoshida H, Nishikawa S (1993) Expression and function of the interleukin 7 receptor in murine lymphocytes. Proc Natl Acad Sci USA 90:9125–9129
Tikhonova AN, Dolgalev I, Hu H, Sivaraj KK, Hoxha E, Cuesta-Dominguez A, Pinho S, Akhmetzyanova I, Gao J, Witkowski M, Guillamot M, Gutkin MC, Zhang Y, Marier C, Diefenbach C, Kousteni S, Heguy A, Zhong H, Fooksman DR, Butler JM, Economides A, Frenette PS, Adams RH, Satija R, Tsirigos A, Aifantis I (2019) The bone marrow microenvironment at single-cell resolution. Nature 569:222–228
Tokoyoda K, Egawa T, Sugiyama T, Choi BI, Nagasawa T (2004) Cellular niches controlling B lymphocyte behavior within bone marrow during development. Immunity 20:707–718
Tsuneto M, Tokoyoda K, Kajikhina E, Hauser AE, Hara T, Tani-ichi S, Ikuta K, Melchers F (2013) B-cell progenitors and precursors change their microenvironment in fetal liver during early development. Stem Cells 31:2800–2812
Watanabe M, Ueno Y, Yajima T, Iwao Y, Tsuchiya M, Ishikawa H, Aiso S, Hibi T, Ishii H (1995) Interleukin 7 is produced by human intestinal epithelial cells and regulates the proliferation of intestinal mucosal lymphocytes. J Clin Invest 95:2945–2953
Whitlock CA, Tidmarsh GF, Muller-Sieburg C, Weissman IL (1987) Bone marrow stromal cell lines with lymphopoietic activity express high levels of a pre-B neoplasia-associated molecule. Cell 48:1009–1021
Whitlock CA, Witte ON (1982) Long-term culture of B lymphocytes and their precursors from murine bone marrow. Proc Natl Acad Sci USA 79:3608–3612
Yokota T, Oritani K, Takahashi I, Ishikawa J, Matsuyama A, Ouchi N, Kihara S, Funahashi T, Tenner AJ, Tomiyama Y, Matsuzawa Y (2000) Adiponectin, a new member of the family of soluble defense collagens, negatively regulates the growth of myelomonocytic progenitors and the functions of macrophages. Blood 96:1723–1732
Zhou BO, Yu H, Yue R, Zhao Z, Rios JJ, Naveiras O, Morrison SJ (2017) Bone marrow adipocytes promote the regeneration of stem cells and haematopoiesis by secreting SCF. Nat Cell Biol 19:891–903
Zhu Y, Cui G, Miyauchi E, Nakanishi Y, Mukohira H, Shimba A, Abe S, Tani-ichi S, Hara T, Nakase H, Chiba T, Sehara-Fujisawa A, Seno H, Ohno H, Ikuta K (2020) Intestinal epithelial cell-derived IL-15 determines local maintenance and maturation of intra-epithelial lymphocytes in the intestine. Int Immunol 32:307–319

Lymph Node Stromal Cells: Diverse Meshwork Structures Weave Functionally Subdivided Niches

Arata Takeuchi, Madoka Ozawa, Guangwei Cui, Koichi Ikuta, and Tomoya Katakai

Abstract Lymph nodes (LNs) are secondary lymphoid organs that function as the first line of defense against invasive foreign substances. Within the LNs, different types of immune cells are strategically localized to induce immune responses efficiently. Such a sophisticated tissue structure is a complex of functionally specialized niches, constructed by a variety of fibroblastic stromal cells. Elucidating the characteristics and functions of the niches and stromal cells will facilitate comprehension of the immune response induced in the LNs. Three recent studies offered novel insights into specialized stromal cells. In our discussion of these surprisingly diverse stromal cells, we will integrate information from these studies to improve knowledge about the structure and niches of LN.

1 Introduction

Lymph nodes (LNs) are situated at critical positions in the lymphatic vascular system and widely present throughout the body to filter lymph fluid exudate. They are also secondary lymphoid organs that function as the first line of defense against invasive pathogens or foreign substances. LNs continuously monitor the lymph fluid and, if antigens are detected, rapidly induce adaptive immune responses. The sophisticated tissue structure in LN is suitable for this purpose; it comprises several distinct areas in which different immune cell types are strategically localized (Qi et al. 2014). The "functional segregation" of immune cells is believed to optimize spatiotemporal regulation of motile behaviors, which induce efficient responses.

Individual LN has one or more basic structures called "compartments (or lobules)," which comprise two major parts, the cortex and medulla (Belisle and Sainte-Marie

A. Takeuchi · M. Ozawa · T. Katakai (✉)
Department of Immunology, Niigata University Graduate School of Medical and Dental Sciences, Niigata 951-8510, Japan
e-mail: katakai@med.niigata-u.ac.jp

G. Cui · K. Ikuta
Laboratory of Immune Regulation, Department of Virus Research, Institute for Frontier Life and Medical Sciences, Kyoto University, Kyoto 606-8507, Japan

T. Nagasawa (ed.), *Bone Marrow Niche*, Current Topics in Microbiology and Immunology 434, https://doi.org/10.1007/978-3-030-86016-5_5

1981; Gretz et al. 1997; Willard-Mack 2006; Sainte-Marie 2010). The cortex is characterized by a dense lymphocyte accumulation in the hemisphere of afferent lymphatic entry. B cells and T cells are localized in different cortical areas; B cells in the superficial cortex forming multiple follicles, and T cells in the deeper layer called the T cell zone (or paracortex). Upon immune response induction, activated B cells start to construct a specialized structure called the germinal center (GC) in the follicle via the help of follicular helper T (Tfh) cells, which plays a vital role in generating humoral immunity (Victora and Nussenzweig 2012). In a single compartment, individual functional areas occupying a certain spatial expanse with the localization of specific immune cells are classified as "subcompartments" (Takeuchi et al. 2018). Subcompartments are tightly associated with unique non-hematopoietic fibroblastic stromal cells (FSCs) that exhibit specialized roles in organizing niches. Each FSC subset produces specific chemokines and cytokines for immune cells' localization and function (Cyster 1999; Mueller and Germain 2009; Katakai 2016). Importantly, the construction and maintenance of subcompartments require continuous interaction between immune cells and stromal cells (Roozendaal and Mebius 2011). For instance, follicular dendritic cells (FDCs), which are FSC subsets in the follicular center or GC light zone, organize subcompartments via the production of CXCL13 that attracts B cells and also Tfh cells (Allen and Cyster 2008; McHeyzer-Williams et al. 2011). Conversely, lymphotoxin (LT)-$\alpha 1\beta 2$ and TNFα produced from B cells is required for the maintenance of FDC function and tissue structure (Matsumoto et al. 1997).

As of 2016, four major FSC subsets associated with cortical subcompartments have been identified, including FDC, TRC (T zone reticular cell), MRC (marginal reticular cell), and CRC (CXCL12-expressing reticular cells) (Chen et al. 1978; Link et al. 2007; Katakai et al. 2008; Bannard et al. 2013; Katakai 2016). Several additional subsets that do not correlate with a particular subcompartment were also proposed (Malhotra et al. 2012; Mionnet et al. 2013; Sitnik et al. 2016). Such a growing heterogeneity of LN stromal cells led to slight confusion in the field. Meanwhile, previous reports show a tendency of bias toward the structure and function of cortical areas due to their noticeable anatomy and dedication to lymphocyte-centered responses. Stromal cell studies have also focused on cortical subsets, and information in other regions such as the medullary region of LN has been limited until recently.

In 2018, three independent studies showed extensive evidence that expands our knowledge on the diversity of LN stromal cells with findings of novel subsets (Table 1). Rodda et al. identified nine clusters in a single-cell RNA sequencing (scRNAseq) of $CD45^{-}CD31^{-}$ non-endothelial stromal cells (Rodda et al. 2018). Huang et al. proposed seven different populations of fibroblastic reticular cells (FRCs) with multiparameter flow cytometry-based clustering, primarily focusing on the medulla (Huang et al. 2018). Based on anatomical examinations using multiple gene reporter mice, Takeuchi et al. reported two novel FSC subsets in the cortex-medulla boundary and medullary cords, and proposed at least six FSC subsets existing in LNs (Takeuchi et al. 2018). Since these studies took different approaches in terms of methodologies and experimental conditions, the proposals of stromal subsets could contain conflicting classifications.

Table 1 Comparative summarization of recent three reports regarding FSC subsets and tissue subcompartments in LN

Subcompartment	2018 Rodda *et al.*	2018 Huang *et al.*	2018 Takeuchi *et al.* + reanalysis	Markers
B cell follicle (B cell zone) SCS-lining, Outer follicle / Follicular mantle	MRC	$Cxcl13^{+}Ccl21^{-}$ MRC	MRC	Madcam1, Tnfsf11, Cxcl13
		$Cxcl13^{-}Ccl21^{+}$ MRC		
Follicular center / GC light zone	FDC	FDC	FDC	Cr1, Cr2, Cxcl13
Inner follicle / GC dark zone	CRC		CRC	Cxcl12
T cell zone (Paracortex) T zone center	$Ccl19^{hi}$ TRC	$Cxcl13^{-}Ccl21^{+}$ TRC	$Ccl19^{hi}$ TRC	Ccl19, Ccl21, Pdpn
	$Cxcl9^{+}$ TRC			Cxcl9, Cxcl10 Activated?
FTI & IFR (CR)	$Ccl19^{lo}$ TRC	$Cxcl13^{+}Ccl21^{-}$ TRC (BRC)	$Ccl19^{lo}$ TRC	Ch25h, Ccl21
DCP		$Cxcl13^{-}Ccl21^{-}$ TRC	DRC	Il7, Cxcl12, Ccl21
Medulla Medullary Cord	$Inmt1^{+}$ SC	$Cxcl13^{-}Ccl21^{+}$ MedRC	$Lepr^{lo}$ MCRC	Cxcl12
		$Cxcl13^{-}Ccl21^{-}$ MedRC	$Lepr^{hi}$ MCRC	Lepr, Cxcl12, Ccl19
	$Nr4a1^{+}$ SC			Nr4a1 Activated?

Here, we comparatively reviewed these reports with a reevaluation of the scRNAseq dataset, and also histochemical confirmation, to enhance knowledge of the structural and functional significance of LN stromal cells. We defined a "stromal cell subset" as a cell population that constructs a subcompartment, i.e., niche for immune cells. We emphasized anatomical viewpoints to identify the stromal subset. Although single-cell and flow cytometry analyses are powerful approaches, positional information was entirely lost during cell isolation, possibly leading to excess subfractions or inseparable closely related subsets. Therefore, careful confirmation of actual microanatomy and cell/molecular distributions was indispensable.

2 Reevaluating a Single-Cell Analysis of LN Stromal Cells

To resolve discrepancies between the "original" single-cell analysis and other reports, we tried to reanalyze the datasets deposited in a database (Rodda et al. 2018). The results could also be incongruous owing to the use of the latest versions of programs, different from that used for the original analysis (Stuart et al. 2019). From the reanalysis, we successfully identified nine clusters (Fig. 1a); each cluster was assigned to the most probable stromal subset based on marker gene expression. Consequently, we obtained similar results for the majority of clusters as the original report, including FDC, MRC, Ccl19^{hi} TRC, Ccl19^{lo} TRC, Cd34^{+} stromal cell (SC), and perivascular cell (PvC) (Fig. 1b, c, and data not shown). Of note, the reanalysis improved some crucial points compared to the original report; (1) MRC fraction was enriched for Madcam1 expression, and (2) Ch25h expression was detected in Ccl19^{lo} TRC fraction (Fig. 1c; described later).

Reanalysis provided a few clusters with clearly different features from the original report (Figs. 1 and 2). The Cxcl9^{+} TRC cluster, proposed as one of the three TRC subsets (Rodda et al. 2018), became part of Ccl19^{hi} TRC fraction, and another Ccl19^{lo}Ccl21^{+} TRC-like cluster emerged in addition to the Ccl19^{lo}Ch25h^{+} TRCs. As the new Ccl19^{lo} cluster appeared to show the characteristics of a specific stromal cell subset in the cortex-medulla boundary (Takeuchi et al. 2018), we assigned this subset to a new cluster (Fig. 1a; described later). Two clusters differentially expressing Nr4a1 and Inmt were initially designated as stromal cells in the medullary cord (MC) (Rodda et al. 2018). However, in our reanalysis, the expression of these genes was comparable in two candidate medullary clusters (Fig. 1a, b). In particular, Inmt expression corresponded well with these clusters. It was also reported that the leptin receptor (Lepr) is highly expressed in medullary cord reticular cells (MCRCs) (Takeuchi et al. 2018). As the two clusters showed enrichment for Lepr expression, we assigned these as MCRCs (Fig. 1a, b). We also noticed that MCRC clusters could be distinguished by the expression level of Lepr, which was well correlated with the histological examinations of LNs (Fig. 2c). Accordingly, we named these Lepr^{hi} MCRC and Lepr^{lo} MCRC, respectively (Fig. 1a; described later). We will not provide details about CD34^{+} SC and PvCs, since these stromal cells are observed in the capsule or surroundings of blood vessels that we do not consider to be subcompartments. Taken together, we propose a current model for subcompartments associated with stromal cell subsets in LN (Fig. 3, Table 1). We will introduce each subset one by one below.

3 B Cell Follicle

Follicles are B cell aggregates located in the outer cortex below the capsule. In this area, three stromal subsets, MRCs, FDCs, and CRCs, provide specialized niches in the outer, center, and inner follicle, respectively. After antigen-specific activation, B

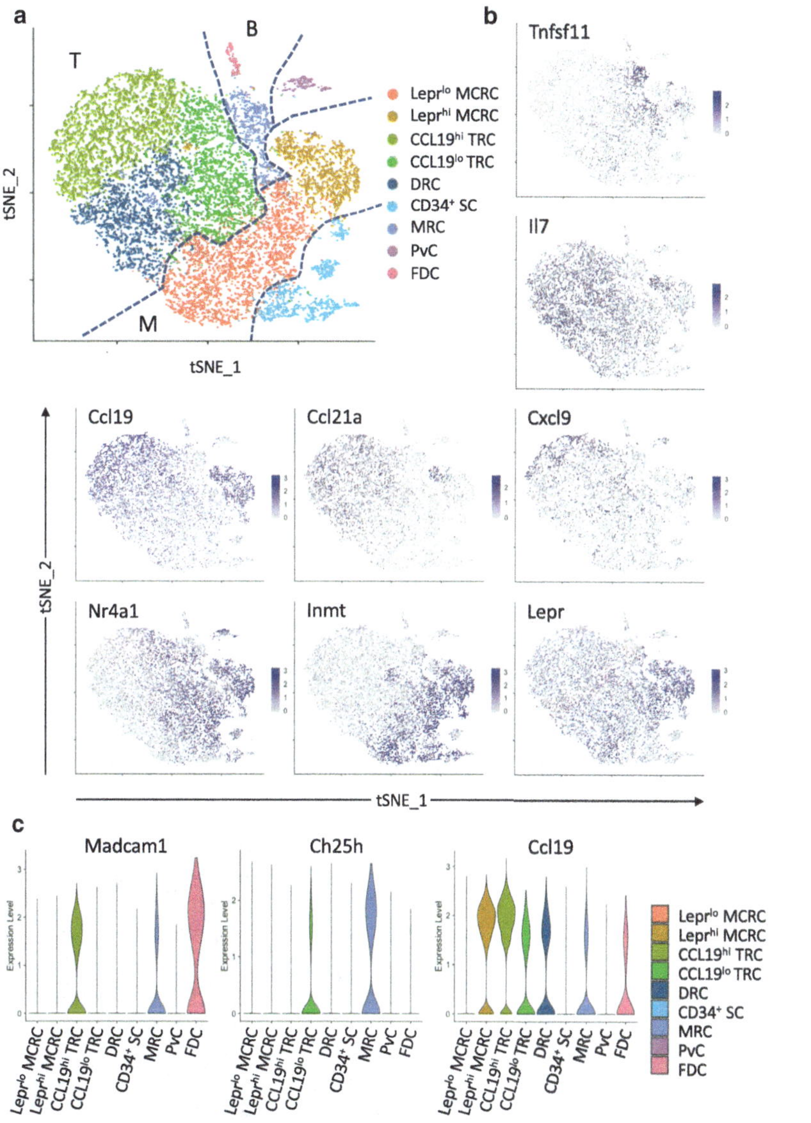

Fig. 1 Identification of 9 clusters by reanalyzing single-cell datasets reported by Rodda et al. 2018. **a** Clustering reanalysis visualized with tSNE. Each dot indicates a single cell that is colored by each cluster categorized. B: B cell zone, T: T cell zone, and M: Medullary cord, respectively. **b** Feature plot of indicated genes. Distribution of gene expression levels is projected on to tSNE plots. **c** Violin plots of Madcam1, Ch25h, and Ccl19 expression in each cluster. Reanalysis of scRNAseq datasets deposited in the GEO database (ID code: GSE112903) was performed using R (version 3.6.2) and Seurat package (version 3.1.0) (Stuart et al. 2019). Default parameters using 10 principle components (PCs) that were determined by the results of JackStrawPlot and ElbowPlot programs set for the analysis. The FindClusters program (resolution = 0.4) finally led to nine clusters, as shown in (**a**)

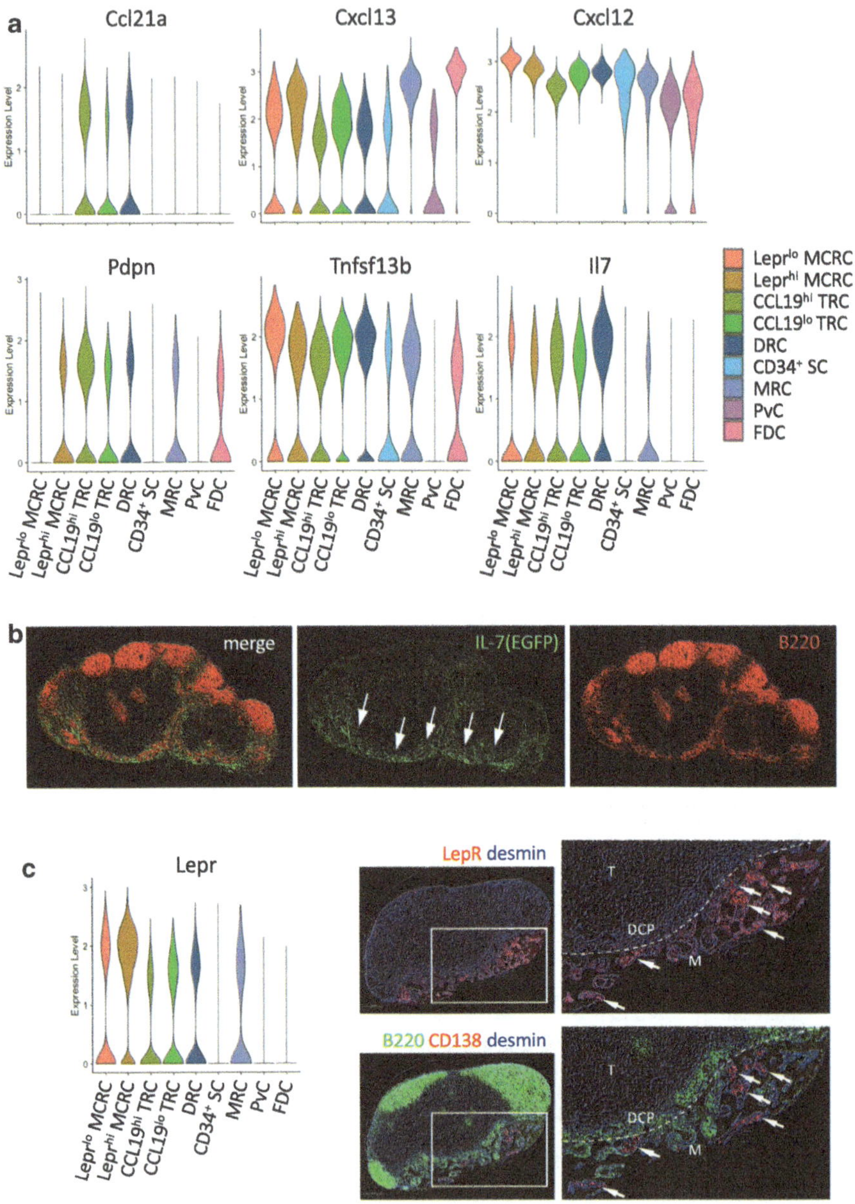

Fig. 2 Novel characteristics in TRC and MCRC subsets. **a** Violin plot of gene expression related to TRC subsets, **b** DRC is a high producer of IL-7. Section of a brachial LN from IL-7 GFP knock-in mouse (Hara et al. 2012) was stained for B220 and desmin. Arrows indicate the DCP area. **c** Expression pattern of Lepr in LN. Violin plot of Lepr expression (left) and immunohistochemistry of LN (right). Section of a cervical LN from C57BL/6 wt mouse was stained for Lepr and desmin (upper right) or B220, CD138, and desmin (lower right). Arrows in higher magnification views indicate the MCs with high Lepr staining colocalized with $CD138^+$ plasma cells

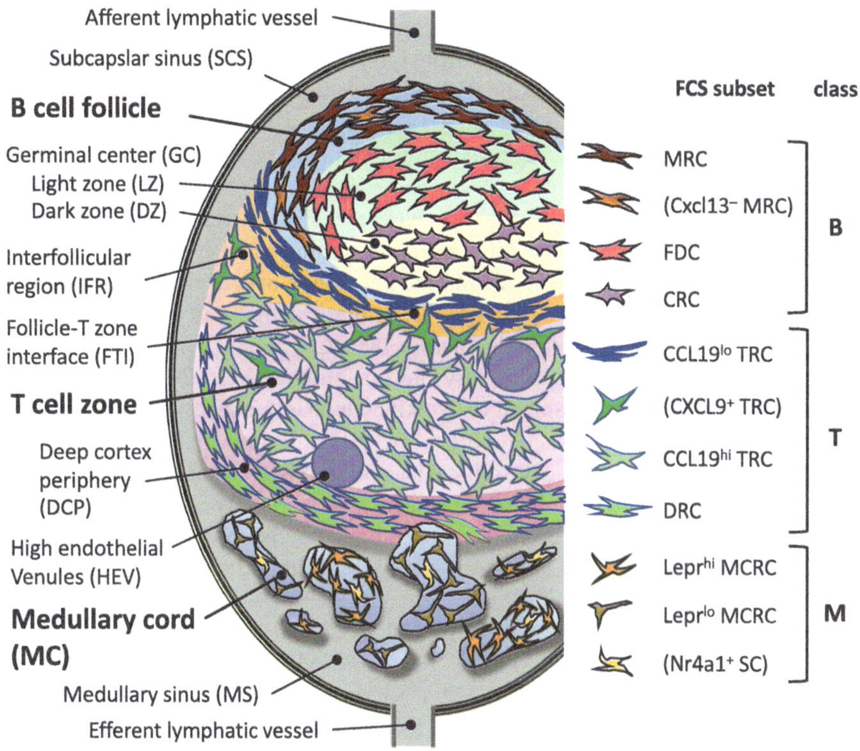

Fig. 3 A proposed model of tissue structure and FSC subsets in LN. Various types of stromal cells construct unique subcompartments, which are highly organized and functionally specialized niches for localized immune cells. The representative basic unit of a single "compartment (lobule)" composed of a single B cell follicle with GC, T cell zone, and medulla is shown. B: B cell zone, T: T cell zone, M: medullary cord

cells start to proliferate, giving rise to the GC in the follicle (Victora and Nussenzweig 2012). Functionally, GC is subdivided into two areas: the dark zone (DZ) where B cells undergo hyperproliferation and somatic hypermutation (SHM) in the B cell receptor (BCR) gene, and the light zone (LZ) in which B cells expressing high-affinity BCR are selected and differentiated into antibody-producing cells (Allen et al. 2007). GC development leads to the production of FDCs and CRCs to support the LZ and DZ, respectively, for the organization of highly specialized niches.

3.1 *Marginal Reticular Cell (MRC)*

MRCs are situated in the outermost part of follicles just below the subcapsular sinus (SCS), where afferent lymphatic inflow occurs in LNs (Katakai et al. 2008). MRCs express CXCL13, MAdCAM-1, and RANKL; expressions of these genes are enriched in the corresponding cluster in single-cell reanalysis (Figs. 1b, c, and 2a). This subset is likely to descend directly from lymphoid tissue organizer (LTo) stromal cells (Hoorweg et al. 2015; Katakai 2012) and have the potential to differentiate into FDCs after proliferation (Jarjour et al. 2014). The specific localization of MRCs at the front line of antigen entry suggests a non-redundant role in forming a niche for a defensive barrier.

Several new findings of MRC function were recently reported; RANKL production in MRC is involved in the differentiation of sinusoidal macrophages (SMs) (Camara et al. 2019). Two types of SMs are positioned in the lymphatic sinuses of LNs; SCS macrophages (SSMs) capture lymph-borne antigens in a non-degradative way and transfer them into follicles across the SCS floor (Carrasco and Batista 2007; Phan et al. 2007), while medullary sinus macrophages (MSMs) exhibit robust phagocytosis and release inflammatory cytokines upon immune responses (Chatziandreou et al. 2017). Most SMs, in particular SSMs, disappeared in FRC-specific RANKL deficiency, suggesting the critical role of RANKL signaling in SM differentiation (Camara et al. 2019). Given that antigens, transferred by SSMs into follicular parenchyma, are further transported by migrating B cells up to FDCs (Phan et al. 2009), MRC production of CXCL13 is likely to be important in this process via facilitating B cell motility.

From anatomical viewpoints, Huang et al. classified FRCs into three subpopulations; MRCs ($BP3^{+}MAdCAM\text{-}1^{+}$), TRCs ($BP3^{+}MAdCAM\text{-}1^{-}$), and medullary reticular cells (MedRCs; $BP3^{-}MAdCAM\text{-}1^{-}$). Using multiparameter flow cytometry-based clustering, they further subdivided MRCs into two populations; $Ccl21^{-}Cxcl13^{+}$ MRC and $Ccl21^{+}Cxcl13^{-}$ MRC (Huang et al. 2018) (Table 1). However, the $Ccl21^{+}Cxcl13^{-}$ subset was undetectable in single-cell reanalysis because almost all the cells in the MRC cluster highly expressed CXCL13. Additionally, MAdCAM-1 expression was detected only in a small fraction of cells (Fig. 1c), even though most MRCs express MAdCAM-1 in immunohistochemistry. As the discrepancies may arise from a difference in the definition of subsets or methodology, it would be essential to confirm the correlation of actual microanatomy and molecular expression.

3.2 Follicular Dendritic Cell (FDC)

FDCs form a dense meshwork in the follicular center or GC-LZ with a unique antigen-presenting property for B cell affinity maturation. They show a distinct transcriptional profile among all FSC subsets in LNs (Rodda et al. 2018). These cells are characterized by the expression of CR1/CD35, CR2/CD21, FcγRIIB/CD32, FcεRII/CD23, MFGE8, MAdCAM-1, and deposition of complement components (Allen and Cyster 2008; Humphrey and Grennan 1982; Tew et al. 1982).

FDCs have been focused on crucial roles in antigen-specific antibody production, presenting immune complexes (IC) to B cells via complement- and Fc- receptors (Papamichail et al. 1975; Tew et al. 2001). Mice deficient in the Cr2 gene show impairment in antigen uptake and reduced responses to low dose or adjuvant-fee antigens, despite retaining GC formation and affinity maturation (Roozendaal and Carroll 2007). They are also deficient in producing serum antibodies and long-lived antibody-forming cells (Barrington et al. 2002). Since internalization via CR1/CR2 keeps antigens intact in FDCs owing to non-degradative recycling and displaying with periodic intervals, this process is assumed to play a crucial role in long-term antigen presentation. However, analysis of FcγRIIB-deficient FDCs demonstrated that this Fc receptor is dispensable for antigen retention (Barrington et al. 2002; Roozendaal and Carroll 2007; Victora and Nussenzweig 2012). Instead, FcγRIIB may be critical in modulating FDC activities associated with trapping IgG-antigen ICs to maintain humoral immunity, recall responses, or both (Barrington et al. 2002; Carroll and Isenman 2012).

Chemokines and cytokines produced by FDCs are necessary to control the GC microenvironment. CXCL13 functions as a chemoattractant for B cells and Tfh cells expressing CXCR5 (Allen and Cyster 2008; McHeyzer-Williams et al. 2011). This chemokine induces the production of LTαβ and TNFα from B cells, which in turn facilitates CXCL13 production by FDCs, driving a positive feedback loop in maintaining follicular structure (Ansel et al. 2000; Cyster et al. 2000). IC-binding onto FDCs enhances IL-6 production, which further promotes B cell SHM and IgG production as well as GC development (Wu et al. 2009). FDCs are also the source of BAFF that is required for the development and survival of B cells (Schneider et al. 1999). Mice deficient in BAFF could not sustain initial GC response due to defects in FDC maturation and lack of the ability to trap and retain ICs (Rahman et al. 2003; Vora et al. 2003). Thus, the engagement of B cells and FDCs induces mutual activation to make a niche for antigen-specific antibody production.

3.3 CXCL12-Expressing Reticular Cells (CRC)

The DZ in GC is a niche for hyperproliferation and SHM of GC-B cells, so-called "centroblasts." Centroblasts express CXCR4, while the ligand CXCL12 is abundant in the DZ (Allen et al. 2004, 2007). CRC was identified as a stromal subset expressing CXCL12 and forming a network in DZ (Bannard et al. 2013). In the primary follicles, CRCs are positioned in close vicinity to the T cell zone suitable for the early stages of GC polarization (Bannard et al. 2013). Although CRCs and FDCs are morphologically similar, CRCs show undetectable or low expression levels in some FDC markers, including CD35 and distinct profile of chemokine in terms of CXCL12 and CXCL13, suggesting that two subsets could provide distinct niches within GCs. CXCL13 production in LZ controls the positioning of GC-B cells, whereas CXCL12 production in DZ is important for attracting CG-B cells from LZ. CXCR4-deficient B cells are unable to access to the DZ, leading to reduced BCR mutation. Interestingly, the proliferation of centroblasts and differentiation into centrocytes do not require DZ access; however, spatial segregation of two niches provides an optimal environment for effective GC responses (Bannard et al. 2013; Rodda et al. 2015).

In single-cell analyses, CRCs could not be detected as an individual cluster (Rodda et al. 2018) (Fig. 1a). FDCs and CRCs are presumably more similar to each other compared to other stromal subsets, even though some of the gene expressions are different. A large number of cells are required to distinguish these subsets in cluster analysis.

4 T Cell Zone

The T cell zone is situated in the deeper part of the cortex, where T cells and DCs accumulate. Naïve T cells actively migrate in this area for scanning antigens presented by DCs and are activated when they detect cognate antigens (Bajenoff et al. 2006; Miller et al. 2002). Therefore, the T cell zone is the center of immune surveillance mediated by T cells. The hallmark of FSCs in this area is CCL21 (Ccl21a) expression that attracts T cells and DCs via CCR7 (Luther et al. 2000; Link et al. 2007; Woolf et al. 2007). These stromal cells construct an elaborate network of extracellular matrix fiber bundle ensheathed by them, the structure known as the "conduit." The conduit functions as a highway for transporting small molecules less than 70 kDa into the deep paracortex (Gretz et al. 1997, 2000; Sixt et al. 2005). Historically, stromal cells in the T cell zone have been termed FRCs due to their characteristic morphology. However, since stromal cells with reticular morphology are found in many other places besides the paracortex (Gretz et al. 1997), FRCs in the T cell zone are more specifically called TRCs (Link et al. 2007).

Recent reports independently suggested that TRCs can be further divided into several subpopulations. Based on single-cell analysis, Rodda et al. proposed three TRC subsets, $CCL19^{hi}$, $CCL19^{lo}$, and $CXCL9^{+}$ TRCs (Rodda et al. 2018). Huang et al. also showed three subsets, $CCL21^{-}CXCL13^{-}$, $CCL21^{-}CXCL13^{+}$, and $CCL21^{+}CXCL13^{-}$ TRCs (Huang et al. 2018). From histological observations, a unique stromal subset expressing CCL21 and CXCL12 was identified in the cortex-medulla boundary (Takeuchi et al. 2018). Here, we discuss the relevance of these reports, together with the reanalysis of the single-cell dataset.

4.1 *CCL19^hi TRC*

$CCL19^{hi}$ TRCs are the most distinct subset expressing typical markers of T zone stromal cells (Rodda et al. 2018) (Figs. 1a and 2a). As $CCL19^{hi}$ TRCs are the primary CCL21 producers, they are likely the equivalent of $CCL21^{+}CXCL13^{-}$ TRC (Huang et al. 2018) (Table 1). CCL21 proteins bound on the surface of TRCs and high endothelial venule through glycosaminoglycans (GAGs) play crucial roles in recruitment and interstitial migration of T cells and DCs (Handel et al. 2005; Woolf et al. 2007). DC migration is also regulated by podoplanin (PDPN), in agreement with the enrichment of PDPN expression in $CCL19^{hi}$ TRCs (Rodda et al. 2018)

(Figs. 1b and 2a). By using C-type lectin receptor CLEC-2 that binds to PDPN, DCs can spread and migrate along with stromal cells (Acton et al. 2014). Additionally, engagement of PDPN and CLEC-2 is necessary to control the TRC network as an optimal niche (Astarita et al. 2015). Under resting conditions, TRCs have less opportunity to encounter CLEC-2^+ mature DCs, thereby PDPN keeps actomyosin contractility in TRCs. When immune response arises, CLEC-2 on DCs causes actomyosin relaxing via PDPN and allows the TRC network to stretch. CLEC-2–PDPN axis also controls ECM production and deposition by TRCs (Martinez et al. 2019).

$CXCL9^+$ TRCs were identified as a cluster in the first single-cell and histological analysis observed in the T cell zone and interfollicular regions (IFRs) (Rodda et al. 2018). They express type 1 IFN, IFN-γ receptors, and some IFN inducible genes. It was also reported that IFN-γ induces CXCL9 production in stromal cells for recruiting memory $CD8^+$ T cells to the periphery of the T cell zone (Sung et al. 2012). Accordingly, $CXCL9^+$ TRCs are suggested to be an activated fraction of $CCL19^{hi}$ TRCs (Rodda et al. 2018). In our reanalysis, $CXCL9^+$ TRCs did not form an independent cluster, but instead, they were mostly included in the $Ccl19^{hi}$ TRC fraction (Fig. 1a, b). Thus, it is still unclear whether $CXCL9^+$ TRCs constitute an individual subset for constructing a specialized niche.

4.2 *$CCL19^{lo}$ TRC*

$CCL19^{lo}$ TRCs express several genes characteristics of both in the T cell zone and follicles, such as CCL21, IL-7, and CXCL13 (Rodda et al. 2018) (Figs. 1b and 2a). Additionally, $CCL19^{lo}$ TRCs highly express BAFF (Tnfsf13b), a cytokine critical for B cell survival. Therefore, it is reasonable that $CCL19^{lo}$ TRCs possibly reside in the IFR and follicle-T zone interface (FTI), making a niche for engaging B and T cells. Previous reports showed that activated B and T cells express EBI2 (Gpr183), a receptor involved in guiding B cells to the outer follicle (Li et al. 2016; Pereira et al. 2009). An EBI2 ligand, 7α,25-dihydroxycholesterol (7α,25-HC) is synthesized by cholesterol-25-hydroxylase (Ch25h) (Hannedouche et al. 2011; Liu et al. 2011; Yi et al. 2012; Li et al. 2016) that is detected in the surrounding of follicles; IFRs and FTI, as well as the SCS-lining (Rodda et al. 2018). Reanalysis of single-cell datasets successfully detected Ch25h expression in $CCL19^{lo}$ TRCs as well as MRCs (Fig. 1c). These findings suggest the perifollicular distribution of $CCL19^{lo}$ TRCs.

Huang et al. identified CXCL13-producing TRC ($CCL21^-CXCL13^+$ TRC) (Huang et al. 2018). Several reports suggest that $CXCL13^+$ FRCs reside in the outer follicle and co-express BAFF, Delta like-4 (DL-4), and Ch25h (Cremasco et al. 2014; Fasnacht et al. 2014; Malhotra et al. 2012; Mionnet et al. 2013; Yi et al. 2012). Huang et al., therefore, proposed to call them "B zone reticular cells (BRCs)" (Huang et al. 2018). However, the anatomical location and gene expression profiles suggest that this population is likely to be $CCL19^{lo}$ TRCs (Table 1). The IFR and FTI correspond to a region named the cortical ridge (CR), where a dense reticular network supports

DC–T–B interactions (Bajenoff et al. 2003; Katakai et al. 2004). It will be interesting to determine if the CR has some relevance to $CCL19^{lo}$ TRCs as a functional subcompartment.

4.3 Deep Cortex Periphery Reticular Cell (DRC)

Based on histological examinations, Sainte-Marie et al. reported a tissue layer with dense reticulum at the boundary of the cortex and medulla, named "deep cortex periphery (DCP)" (Belisle et al. 1982; Sainte-Marie 2010). We recently found that B cells accumulate in the DCP in a CCL21- and CXCL12-dependent manner (Takeuchi et al. 2018). Stromal cells in the DCP showed unique properties to form a dense network expressing both CCL21 and CXCL12, allowing us to define them as a new subset named "DCP reticular cells (DRCs)." Interestingly, B and T cells are intermingled in this area, in marked contrast to the follicles. B cells in the DCP migrate at a slower velocity and higher turning angle than that in the follicles, suggesting that DRCs provide a specialized microenvironment. The optimal development of DCP involves B cells, because DCP formation was attenuated in B cell-deficient mice but restored by bone marrow transfer. These results suggest that DRCs require signals from B cells to construct the DCP (Takeuchi et al. 2018).

In the first single-cell analysis, no prominent cluster corresponding to DRCs was detected (Rodda et al. 2018). However, in our reanalysis, another $CCL19^{lo}$ cluster was found in addition to $CCL19^{lo}Ch25h^{+}$ TRCs (Fig. 1a, c). We considered that the new cluster represented DRCs because of high CXCL12 and CCL21 expression (Fig. 2a). Meanwhile, this cluster also showed the highest IL-7 expression in all stromal clusters (Figs. 1b and 2a). Immunohistochemical analysis using IL-7 reporter mice demonstrated that IL-7 expression was prominent in the DCP compared with other parts of LN (Fig. 2b), which convinced us that this cluster could be assigned as DRCs. DRC cluster shared some features with $CCL19^{lo}$ TRCs, such as the expression of BAFF (Fig. 2a). These suggest that the DCP area supported by DRCs is a niche suitable for lymphocyte survival. Contrarily, Ch25h expression was weak in DRCs (Fig. 1c), suggesting that DRCs and $CCL19^{lo}$ TRCs are likely to differ in function during immune responses.

5 Medullary Cord

Lymphocytes are highly concentrated in the cortex, whereas the medulla appears to have relatively low cell density due to the abundance of lymphatic sinus called the medullary sinus (MS). In the medulla, mesenchymal cells are mostly assembled to form the medullary cords (MCs); these are sheath-like parenchymal structures surrounding blood vessels (Ohtani and Ohtani 2008). It was considered that medullary FSCs were distinct from other subsets as this region harbors completely different types of immune cells, such as macrophages, NK cells, and plasma cells (Gray and Cyster 2012; Hargreaves et al. 2001; Kastenmuller et al. 2012; Katakai et al. 2004). However, their precise nature had remained unclear until recently.

5.1 *Medullary Cord Reticular Cell (MCRC)*

Huang et al. intensively characterized reticular stromal cells in the MCs (MedRCs) as a new subset with the ability to regulate plasma cell homeostasis (Huang et al. 2018). MedRCs differ from TRCs in terms of lower expressions of CCL21, CXCL13, CCL19, and IL-7, but a higher IL-6. MedRCs are the primary source of IL-6, BAFF, and CXCL12, besides also producing APRIL (Cyster 2005; Mohr et al. 2009; Abe et al. 2012). Among these, IL-6 is especially important for the differentiation and survival of plasma cells, since the inhibition of IL-6, but not BAFF, APRIL, and CXCL12, block those events. On the other hand, homing and localization of plasma cells in the MCs depends on the CXCL12-CXCR4 axis (Hargreaves et al. 2001; Cyster 2005). MedRCs express a higher CXCL12 but lower CCL21 and CXCL13 than other subsets (Fig. 2a). Such a profile of chemokine expression is likely to attract plasma cells and macrophages expressing high CXCR4 but low CCR7 and CXCR5.

Huang et al. suggested that MedRCs are composed of two subpopulations, $CCL21^{+}CXCL13^{-}$ and $CCX21^{-}CXCL13^{+}$ cells (Huang et al. 2018). From a single-cell analysis, Rodda et al. also showed two subsets of medullary stromal cells; $Inmt^{+}$ and $Nr4a1^{+}$ SCs. They suggested a possibility that $Nr4a1^{+}$ SCs are the activated phenotype of other subsets, including $Inmt^{+}$ SCs (Rodda et al. 2018). Interestingly, in the reanalysis of a single-cell dataset, Inmt and Nr4a1 were broadly expressed in two medullary clusters rather than enriched in either of the clusters (Fig. 1a, b). We have independently identified an FSC subset in the MC; the MCRCs categorized as $CXCL12^{hi}CCL21^{low}Lepr^{+}$ cells (Takeuchi et al. 2018). Of note, the two medullary clusters corresponded with $Lepr^{hi}$ and $Lepr^{lo}$ fractions (Figs. 1b and 2c). Histological observations indicated that the MCs could be roughly separated into $Lepr^{hi}$ and $Lepr^{lo}$ areas, which appeared to correlate with the localization of plasma cells and B cells, respectively (Fig. 2c). Therefore, two types of MCRCs possibly represent different niches within the MCs.

6 Concluding Remarks

Integrating the three core reports and other previous knowledge led to a comprehensive picture of LN architecture founded upon stromal cell diversity. We thereby propose the latest model of the stromal structure, in which at least eight FSC subsets tightly associated with subcompartments construct a functionally organized "niche complex" (Table 1, Fig. 3). Additionally, closely related subsets in major areas of

LN can be classified into three upper categories; classes "B (B cell follicle)," "T (T cell zone)," and "M (medulla)."

Within the B cell follicles, the "class B" members MRCs, FDCs, and CRCs support distinct follicular niches. The functions of MRCs are gradually uncovered recently, in particular, the control of niches in the antigenic entry site via RANKL expression. FDCs constitute a prominent structure in the follicular center and GC-LZ, which is reflected by a discrete cluster expressing a variety of specific genes in single-cell analysis. In contrast, even though CRCs are histologically apparent in the inner follicle or GC-DZ, the corresponding cluster is currently undetected in single-cell analysis, probably due to the limited cell number of the datasets.

The "class T" includes FSC subsets in the T cell zone, $CCL19^{hi}$ TRCs, $CCL19^{lo}$ TRCs, and DRCs. The hallmark of these subsets is CCL21 expression. $CCL19^{hi}$ TRCs constitute the center body of the T cell zone by highly producing CCL19 as well as CCL21. In concert with CLEC-2^{+} DCs, PDPN highly expressed in this subset is suggested to control the scale of the stromal network. $CCL19^{lo}$ TRCs are likely the stromal component of the IFR and FTI (CR), providing a niche for B and T cell engagement. DRCs support the DCP area associated with B cell accumulation at the periphery of the deep T cell zone. Although DRCs highly express various survival factors for lymphocytes, it is still unclear why B cells need to be attracted to the opposite side of the follicles near the medulla.

Stromal cells of the medulla could be categorized as "class M" subsets providing a niche for plasma cells and medullary macrophages. Two types of FSCs possibly constitute distinct areas within the MC; $Lepr^{hi}CCL19^{+}$ MCRCs and $Lepr^{lo}$ MCRCs appear to correlate with the accumulations of plasma cells and B cells, respectively. There are some discrepancies between reports; however, the MCs are presumably composed of more than one FSC subset.

The subset composition of LN stromal cells in its entirety is surprisingly complicated and diverse to form functionally different subcompartments. Such a stromal cell diversity is probably indispensable to control elaborate and flexible immune responses. We speculate that the feature of each subset is not uniform nor stable but has overlaps and interchangeable property with neighboring populations. We, therefore, assume that it is possible to discover more FSC types or subsets with intermediate property, especially in the boundary of subcompartments. It is also possible that the condition of stromal cells, subset composition, or both, could be readily changed during immune responses or local stimuli. Overall, further in-depth analysis will be necessary to obtain a universal picture of the diversity and plasticity of LN FSCs.

Acknowledgements This work was funded by a Grant-in-Aid for Scientific Research (C) (A. Takeuchi, 19K07603) and a Grant-in-Aid for Challenging Exploratory Research (T. Katakai, 16K15287) from The Ministry of Education, Culture, Sports, Science and Technology of Japan.

Author's note: Due to a significant delay in the publication process caused by the COVID-19 pandemic, the content of this paper is as of April 2020.

References

Abe J, Ueha S, Yoneyama H, Shono Y, Kurachi M, Goto A, Fukayama M, Tomura M, Kakimi K, Matsushima K (2012) B cells regulate antibody responses through the medullary remodeling of inflamed lymph nodes. Int Immunol 24:17–27

Acton SE, Farrugia AJ, Astarita JL, Mourao-Sa D, Jenkins RP, Nye E, Hooper S, van Blijswijk J, Rogers NC, Snelgrove KJ, Rosewell I, Moita LF, Stamp G, Turley SJ, Sahai E, Reis e Sousa C (2014) Dendritic cells control fibroblastic reticular network tension and lymph node expansion. Nature 514:498–502

Allen CD, Ansel KM, Low C, Lesley R, Tamamura H, Fujii N, Cyster JG (2004) Germinal center dark and light zone organization is mediated by CXCR4 and CXCR5. Nat Immunol 5:943–952

Allen CD, Cyster JG (2008) Follicular dendritic cell networks of primary follicles and germinal centers: phenotype and function. Semin Immunol 20:14–25

Allen CD, Okada T, Cyster JG (2007) Germinal-center organization and cellular dynamics. Immunity 27:190–202

Ansel KM, Ngo VN, Hyman PL, Luther SA, Forster R, Sedgwick JD, Browning JL, Lipp M, Cyster JG (2000) A chemokine-driven positive feedback loop organizes lymphoid follicles. Nature 406:309–314

Astarita JL, Cremasco V, Fu J, Darnell MC, Peck JR, Nieves-Bonilla JM, Song K, Kondo Y, Woodruff MC, Gogineni A, Onder L, Ludewig B, Weimer RM, Carroll MC, Mooney DJ, Xia L, Turley SJ (2015) The CLEC-2-podoplanin axis controls the contractility of fibroblastic reticular cells and lymph node microarchitecture. Nat Immunol 16:75–84

Bajenoff M, Egen JG, Koo LY, Laugier JP, Brau F, Glaichenhaus N, Germain RN (2006) Stromal cell networks regulate lymphocyte entry, migration, and territoriality in lymph nodes. Immunity 25:989–1001

Bajenoff M, Granjeaud S, Guerder S (2003) The strategy of T cell antigen-presenting cell encounter in antigen-draining lymph nodes revealed by imaging of initial T cell activation. J Exp Med 198:715–724

Bannard O, Horton RM, Allen CD, An J, Nagasawa T, Cyster JG (2013) Germinal center centroblasts transition to a centrocyte phenotype according to a timed program and depend on the dark zone for effective selection. Immunity 39:912–924

Barrington RA, Pozdnyakova O, Zafari MR, Benjamin CD, Carroll MC (2002) B lymphocyte memory: role of stromal cell complement and FcgammaRIIB receptors. J Exp Med 196:1189–1199

Belisle C, Sainte-Marie G (1981) Topography of the deep cortex of the lymph nodes of various mammalian species. Anat Rec 201:553–561

Belisle C, Sainte-Marie G, Peng FS (1982) Tridimensional study of the deep cortex of the rat lymph node. VI. The deep cortex units of the germ-free rat. Am J Pathol 107:70–78

Camara A, Cordeiro OG, Alloush F, Sponsel J, Chypre M, Onder L, Asano K, Tanaka M, Yagita H, Ludewig B, Flacher V, Mueller CG (2019) Lymph node mesenchymal and endothelial stromal cells cooperate via the RANK-RANKL cytokine axis to shape the sinusoidal macrophage niche. Immunity 50:1467–1481 e1466

Carrasco YR, Batista FD (2007) B cells acquire particulate antigen in a macrophage-rich area at the boundary between the follicle and the subcapsular sinus of the lymph node. Immunity 27:160–171

Carroll MC, Isenman DE (2012) Regulation of humoral immunity by complement. Immunity 37:199–207

Chatziandreou N, Farsakoglu Y, Palomino-Segura M, D'Antuono R, Pizzagalli DU, Sallusto F, Lukacs-Kornek V, Uguccioni M, Corti D, Turley SJ, Lanzavecchia A, Carroll MC, Gonzalez SF (2017) Macrophage death following influenza vaccination initiates the inflammatory response that promotes dendritic cell function in the draining lymph node. Cell Rep 18:2427–2440

Chen LL, Frank AM, Adams JC, Steinman RM (1978) Distribution of horseradish peroxidase (HRP)-anti-HRP immune complexes in mouse spleen with special reference to follicular dendritic cells. J Cell Biol 79:184–199

Cremasco V, Woodruff MC, Onder L, Cupovic J, Nieves-Bonilla JM, Schildberg FA, Chang J, Cremasco F, Harvey CJ, Wucherpfennig K, Ludewig B, Carroll MC, Turley SJ (2014) B cell homeostasis and follicle confines are governed by fibroblastic reticular cells. Nat Immunol 15:973–981

Cyster JG (1999) Chemokines and cell migration in secondary lymphoid organs. Science 286:2098–2102

Cyster JG (2005) Chemokines, sphingosine-1-phosphate, and cell migration in secondary lymphoid organs. Annu Rev Immunol 23:127–159

Cyster JG, Ansel KM, Reif K, Ekland EH, Hyman PL, Tang HL, Luther SA, Ngo VN (2000) Follicular stromal cells and lymphocyte homing to follicles. Immunol Rev 176:181–193

Fasnacht N, Huang HY, Koch U, Favre S, Auderset F, Chai Q, Onder L, Kallert S, Pinschewer DD, MacDonald HR, Tacchini-Cottier F, Ludewig B, Luther SA, Radtke F (2014) Specific fibroblastic niches in secondary lymphoid organs orchestrate distinct Notch-regulated immune responses. J Exp Med 211:2265–2279

Gray EE, Cyster JG (2012) Lymph node macrophages. J Innate Immun 4:424–436

Gretz JE, Anderson AO, Shaw S (1997) Cords, channels, corridors and conduits: critical architectural elements facilitating cell interactions in the lymph node cortex. Immunol Rev 156:11–24

Gretz JE, Norbury CC, Anderson AO, Proudfoot AE, Shaw S (2000) Lymph-borne chemokines and other low molecular weight molecules reach high endothelial venules via specialized conduits while a functional barrier limits access to the lymphocyte microenvironments in lymph node cortex. J Exp Med 192:1425–1440

Handel TM, Johnson Z, Crown SE, Lau EK, Proudfoot AE (2005) Regulation of protein function by glycosaminoglycans–as exemplified by chemokines. Annu Rev Biochem 74:385–410

Hannedouche S, Zhang J, Yi T, Shen W, Nguyen D, Pereira JP, Guerini D, Baumgarten BU, Roggo S, Wen B, Knochenmuss R, Noel S, Gessier F, Kelly LM, Vanek M, Laurent S, Preuss I, Miault C, Christen I, Karuna R, Li W, Koo DI, Suply T, Schmedt C, Peters EC, Falchetto R, Katopodis A, Spanka C, Roy MO, Detheux M, Chen YA, Schultz PG, Cho CY, Seuwen K, Cyster JG, Sailer AW (2011) Oxysterols direct immune cell migration via EBI2. Nature 475:524–527

Hara T, Shitara S, Imai K, Miyachi H, Kitano S, Yao H, Tani-ichi S, Ikuta K (2012) Identification of IL-7-producing cells in primary and secondary lymphoid organs using IL-7-GFP knock-in mice. J Immunol 189:1577–1584

Hargreaves DC, Hyman PL, Lu TT, Ngo VN, Bidgol A, Suzuki G, Zou YR, Littman DR, Cyster JG (2001) A coordinated change in chemokine responsiveness guides plasma cell movements. J Exp Med 194:45–56

Hoorweg K, Narang P, Li Z, Thuery A, Papazian N, Withers DR, Coles MC, Cupedo T (2015) A stromal cell niche for human and mouse type 3 innate lymphoid cells. J Immunol 195:4257–4263

Huang HY, Rivas-Caicedo A, Renevey F, Cannelle H, Peranzoni E, Scarpellino L, Hardie DL, Pommier A, Schaeuble K, Favre S, Vogt TK, Arenzana-Seisdedos F, Schneider P, Buckley CD, Donnadieu E, Luther SA (2018) Identification of a new subset of lymph node stromal cells involved in regulating plasma cell homeostasis. Proc Natl Acad Sci U S A 115:E6826–E6835

Humphrey JH, Grennan D (1982) Isolation and properties of spleen follicular dendritic cells. Adv Exp Med Biol 149:823–827

Jarjour M, Jorquera A, Mondor I, Wienert S, Narang P, Coles MC, Klauschen F, Bajenoff M (2014) Fate mapping reveals origin and dynamics of lymph node follicular dendritic cells. J Exp Med 211:1109–1122

Kastenmuller W, Torabi-Parizi P, Subramanian N, Lammermann T, Germain RN (2012) A spatially-organized multicellular innate immune response in lymph nodes limits systemic pathogen spread. Cell 150:1235–1248

Katakai T (2016) Stromal cells in secondary lymphoid organs. Encyclopedia of Immunobiology 3:473

Katakai T (2012) Marginal reticular cells: a stromal subset directly descended from the lymphoid tissue organizer. Front Immunol 3:200

Katakai T, Hara T, Lee JH, Gonda H, Sugai M, Shimizu A (2004) A novel reticular stromal structure in lymph node cortex: an immuno-platform for interactions among dendritic cells, T cells and B cells. Int Immunol 16:1133–1142

Katakai T, Suto H, Sugai M, Gonda H, Togawa A, Suematsu S, Ebisuno Y, Katagiri K, Kinashi T, Shimizu A (2008) Organizer-like reticular stromal cell layer common to adult secondary lymphoid organs. J Immunol 181:6189–6200

Li J, Lu E, Yi T, Cyster JG (2016) EBI2 augments Tfh cell fate by promoting interaction with IL-2-quenching dendritic cells. Nature 533:110–114

Link A, Vogt TK, Favre S, Britschgi MR, Acha-Orbea H, Hinz B, Cyster JG, Luther SA (2007) Fibroblastic reticular cells in lymph nodes regulate the homeostasis of naive T cells. Nat Immunol 8:1255–1265

Liu C, Yang XV, Wu J, Kuei C, Mani NS, Zhang L, Yu J, Sutton SW, Qin N, Banie H, Karlsson L, Sun S, Lovenberg TW (2011) Oxysterols direct B-cell migration through EBI2. Nature 475:519–523

Luther SA, Tang HL, Hyman PL, Farr AG, Cyster JG (2000) Coexpression of the chemokines ELC and SLC by T zone stromal cells and deletion of the ELC gene in the plt/plt mouse. Proc Natl Acad Sci U S A 97:12694–12699

Malhotra D, Fletcher AL, Astarita J, Lukacs-Kornek V, Tayalia P, Gonzalez SF, Elpek KG, Chang SK, Knoblich K, Hemler ME, Brenner MB, Carroll MC, Mooney DJ, Turley SJ, Immunological Genome Project C (2012) Transcriptional profiling of stroma from inflamed and resting lymph nodes defines immunological hallmarks. Nat Immunol 13:499–510

Martinez VG, Pankova V, Krasny L, Singh T, Makris S, White IJ, Benjamin AC, Dertschnig S, Horsnell HL, Kriston-Vizi J, Burden JJ, Huang PH, Tape CJ, Acton SE (2019) Fibroblastic reticular cells control conduit matrix deposition during lymph node expansion. Cell Rep 29:2810–2822 e2815

Matsumoto M, Fu YX, Molina H, Huang G, Kim J, Thomas DA, Nahm MH, Chaplin DD (1997) Distinct roles of lymphotoxin alpha and the type I tumor necrosis factor (TNF) receptor in the establishment of follicular dendritic cells from non-bone marrow-derived cells. J Exp Med 186:1997–2004

McHeyzer-Williams M, Okitsu S, Wang N, McHeyzer-Williams L (2011) Molecular programming of B cell memory. Nat Rev Immunol 12:24–34

Miller MJ, Wei SH, Parker I, Cahalan MD (2002) Two-photon imaging of lymphocyte motility and antigen response in intact lymph node. Science 296:1869–1873

Mionnet C, Mondor I, Jorquera A, Loosveld M, Maurizio J, Arcangeli ML, Ruddle NH, Nowak J, Aurrand-Lions M, Luche H, Bajenoff M (2013) Identification of a new stromal cell type involved in the regulation of inflamed B cell follicles. PLoS Biol 11:e1001672

Mohr E, Serre K, Manz RA, Cunningham AF, Khan M, Hardie DL, Bird R, MacLennan IC (2009) Dendritic cells and monocyte/macrophages that create the IL-6/APRIL-rich lymph node microenvironments where plasmablasts mature. J Immunol 182:2113–2123

Mueller SN, Germain RN (2009) Stromal cell contributions to the homeostasis and functionality of the immune system. Nat Rev Immunol 9:618–629

Ohtani O, Ohtani Y (2008) Structure and function of rat lymph nodes. Arch Histol Cytol 71:69–76

Papamichail M, Gutierrez C, Embling P, Johnson P, Holborow EJ, Pepys MB (1975) Complement dependence of localisation of aggregated IgG in germinal centres. Scand J Immunol 4:343–347

Pereira JP, Kelly LM, Xu Y, Cyster JG (2009) EBI2 mediates B cell segregation between the outer and centre follicle. Nature 460:1122–1126

Phan TG, Green JA, Gray EE, Xu Y, Cyster JG (2009) Immune complex relay by subcapsular sinus macrophages and noncognate B cells drives antibody affinity maturation. Nat Immunol 10:786–793

Phan TG, Grigorova I, Okada T, Cyster JG (2007) Subcapsular encounter and complement-dependent transport of immune complexes by lymph node B cells. Nat Immunol 8:992–1000

Qi H, Kastenmuller W, Germain RN (2014) Spatiotemporal basis of innate and adaptive immunity in secondary lymphoid tissue. Annu Rev Cell Dev Biol 30:141–167

Rahman ZS, Rao SP, Kalled SL, Manser T (2003) Normal induction but attenuated progression of germinal center responses in BAFF and BAFF-R signaling-deficient mice. J Exp Med 198:1157–1169

Rodda LB, Bannard O, Ludewig B, Nagasawa T, Cyster JG (2015) Phenotypic and morphological properties of germinal center dark zone Cxcl12-expressing reticular cells. J Immunol 195:4781–4791

Rodda LB, Lu E, Bennett ML, Sokol CL, Wang X, Luther SA, Barres BA, Luster AD, Ye CJ, Cyster JG (2018) Single-cell RNA sequencing of lymph node stromal cells reveals niche-associated heterogeneity. Immunity 48:1014–1028 e1016

Roozendaal R, Carroll MC (2007) Complement receptors CD21 and CD35 in humoral immunity. Immunol Rev 219:157–166

Roozendaal R, Mebius RE (2011) Stromal cell-immune cell interactions. Annu Rev Immunol 29:23–43

Sainte-Marie G (2010) The lymph node revisited: development, morphology, functioning, and role in triggering primary immune responses. Anat Rec (hoboken) 293:320–337

Schneider P, MacKay F, Steiner V, Hofmann K, Bodmer JL, Holler N, Ambrose C, Lawton P, Bixler S, Acha-Orbea H, Valmori D, Romero P, Werner-Favre C, Zubler RH, Browning JL, Tschopp J (1999) BAFF, a novel ligand of the tumor necrosis factor family, stimulates B cell growth. J Exp Med 189:1747–1756

Sitnik KM, Wendland K, Weishaupt H, Uronen-Hansson H, White AJ, Anderson G, Kotarsky K, Agace WW (2016) Context-dependent development of lymphoid stroma from adult CD34(+) adventitial progenitors. Cell Rep 14:2375–2388

Sixt M, Kanazawa N, Selg M, Samson T, Roos G, Reinhardt DP, Pabst R, Lutz MB, Sorokin L (2005) The conduit system transports soluble antigens from the afferent lymph to resident dendritic cells in the T cell area of the lymph node. Immunity 22:19–29

Stuart T, Butler A, Hoffman P, Hafemeister C, Papalexi E, Mauck WM, Hao Y, Stoeckius M, Smibert P, Satija R (2019) Comprehensive integration of single-cell data. Cell 177:1888–1902 e1821

Sung JH, Zhang H, Moseman EA, Alvarez D, Iannacone M, Henrickson SE, de la Torre JC, Groom JR, Luster AD, von Andrian UH (2012) Chemokine guidance of central memory T cells is critical for antiviral recall responses in lymph nodes. Cell 150:1249–1263

Takeuchi A, Ozawa M, Kanda Y, Kozai M, Ohigashi I, Kurosawa Y, Rahman MA, Kawamura T, Shichida Y, Umemoto E, Miyasaka M, Ludewig B, Takahama Y, Nagasawa T, Katakai T (2018) A distinct subset of fibroblastic stromal cells constitutes the cortex-medulla boundary subcompartment of the lymph node. Front Immunol 9:2196

Tew JG, Thorbecke GJ, Steinman RM (1982) Dendritic cells in the immune response: characteristics and recommended nomenclature (A report from the reticuloendothelial society committee on nomenclature). J Reticuloendothel Soc 31:371–380

Tew JG, Wu J, Fakher M, Szakal AK, Qin D (2001) Follicular dendritic cells: beyond the necessity of T-cell help. Trends Immunol 22:361–367

Victora GD, Nussenzweig MC (2012) Germinal centers. Annu Rev Immunol 30:429–457

Vora KA, Wang LC, Rao SP, Liu ZY, Majeau GR, Cutler AH, Hochman PS, Scott ML, Kalled SL (2003) Cutting edge: germinal centers formed in the absence of B cell-activating factor belonging to the TNF family exhibit impaired maturation and function. J Immunol 171:547–551

Willard-Mack CL (2006) Normal structure, function, and histology of lymph nodes. Toxicol Pathol 34:409–424

Woolf E, Grigorova I, Sagiv A, Grabovsky V, Feigelson SW, Shulman Z, Hartmann T, Sixt M, Cyster JG, Alon R (2007) Lymph node chemokines promote sustained T lymphocyte motility without triggering stable integrin adhesiveness in the absence of shear forces. Nat Immunol 8:1076–1085

Wu Y, El Shikh ME, El Sayed RM, Best AM, Szakal AK, Tew JG (2009) IL-6 produced by immune complex-activated follicular dendritic cells promotes germinal center reactions, IgG responses and somatic hypermutation. Int Immunol 21:745–756

Yi T, Wang X, Kelly LM, An J, Xu Y, Sailer AW, Gustafsson JA, Russell DW, Cyster JG (2012) Oxysterol gradient generation by lymphoid stromal cells guides activated B cell movement during humoral responses. Immunity 37:535–548

Immune Niche Within the Peritoneal Cavity

Yasutaka Okabe

Abstract There are numbers of leukocytes present in peritoneal cavity, not only protecting body cavity from infection but also contributing to peripheral immunity including natural antibody production in circulation. The peritoneal leukocytes compose unique immune compartment, the functions of which cannot be replaced by other lymphoid organs. Atypical lymphoid clusters, called "milky spots", that are located in visceral adipose tissue omentum have the privilege of immune niche in terms of differentiation, recruitment, and activation of peritoneal immunity, yet mechanisms underlying the regulation are underexplored. In this review, I discuss the emerging views of peritoneal immune system in the contexts of its development, organization, and functions.

1 Introduction

Body cavities are the fluid-filled spaces where visceral organs and other structures are composed. In human, there are mainly two body cavities exist: ventral cavity is at the anterior of the trunk including the peritoneal and pleural cavity, and dorsal cavity accommodates the brain and spinal cord (Isaza-Restrepo et al. 2018).

Peritoneal cavity is the largest body cavity surrounded by mesothelial layers and accommodate stomach, spleen, intestines, pancreas, reproductive organs, and so on. It contains small volume of peritoneal fluid (50–100 ml in healthy human) which lubricate the surface of visceral tissues (Healy and Reznek 1998). The volume of peritoneal fluid increases (more than litters) during pathological conditions such as liver cirrhosis, cancer, and pancreatitis. This is caused by the increase of the capillary pressure in the abdominal viscera by which fluid enters the peritoneal cavity, a condition called ascites (Szender et al. 2017). Thus, the peritoneal environment changes dynamically upon various physiological and pathological conditions.

Y. Okabe (✉)
Laboratory of Immune Homeostasis, WPI Immunology Frontier Research Center Osaka University, 3-1 Yamada-oka, Suita 565-0871, Osaka, Japan
e-mail: yokabe@ifrec.osaka-u.ac.jp

T. Nagasawa (ed.), *Bone Marrow Niche*, Current Topics in Microbiology and Immunology 434, https://doi.org/10.1007/978-3-030-86016-5_6

Peritoneal cavity is normally sterile and infection in the body cavities causes a life-threatening emergency that needs prompt medical treatment. There are substantial numbers of immune cells present in the peritoneal cavity, protecting the cavity from infection and injury. The immune composition of the peritoneal cavity is quite distinct from that of the other lymphoid organs. For instance, resident macrophages, which account for about a half of the peritoneal leukocytes, exhibit highly specialized functions as a result of their adaptation to peritoneal environment. Peritoneal macrophages play critical roles in the protection of the cavity through the elimination of invaded pathogens, programmed for "silent" clearance of apoptotic cells (Roberts et al. 2017), and maintenance of the population of peritoneal lymphocytes through production of CXC-chemokine ligand 13 (CXCL13) (Ansel et al. 2002). Furthermore, recent study elucidates that the peritoneal cavity serves as the reservoir of mature macrophages that enable prompt migration into the injury sites of visceral tissue parenchyma for the promotion of tissue repair responses (Wang and Kubes 2016).

The peritoneal leukocytes also contain lymphocytes, eosinophils, monocytes, dendritic cells, innate lymphoid cells (ILCs), and so on. Especially, peritoneal cavity harbors the unique B lymphocyte subset, B-1 cells, which dominate approximately 50% of B lymphocytes in body cavities, whereas the identification of B-1 cells outside of the body cavities is challenging because of their low frequency (~1% in lymph node and spleen) (Baumgarth 2011). B-1 cells are distinguished from conventional B (B-2) cells by the expression of a BCR repertoire enriched for highly poly-specific receptors that bind to both self-antigens and microbial antigens, and they are involved mainly in T cell-independent and innate-like immune responses (Baumgarth 2016). Whereas B-1 cells residing in the peritoneal cavity do not spontaneously generate antibodies, they can be readily migrate to the periphery (e.g., spleen and intestine) where they differentiate into IgM or IgA secreting cells. Taken together, peritoneal immunity is not only unique immune system for body cavity protection but it is also associated with visceral tissue homeostasis and peripheral immunity. In this review, I summarize the current understanding about development, organization, and functions of the peritoneal immune system and its associated-lymphoid tissues.

2 Omentum Functions as "Police of the Abdomen" for Protection from Abdominal Infection

A visceral tissue, omentum, plays central roles in the development, organization, and functions of peritoneal immunity (Wilkosz et al. 2005). Omentum is a sheet of visceral adipose fat connecting with the spleen, stomach, pancreas, and colon (Di Nicola 2019). Anatomically, two types of omenta are referred (Yoo et al. 2007): the greater omentum is a large apron-like fold of visceral adipose attached along the whole length of the greater curvature of the stomach, while the lesser omentum hangs down from the lesser curvature. The omentum tissue in human can store large amounts of adipose tissue; the size is from 300 to 2,000 g with a surface area of

300 to 1,500 cm^2 and its size can expand under the situations such as obesity and inflammation (Di Nicola 2019). Thus, the omentum is of great interest in the context of obesity-related metabolic disease (O'Connell et al. 2010). In contrast, omentum in mouse is usually found as a relatively small strip of fat attached stomach, although it is structurally similar to human omentum.

Unlike typical adipose tissues, the omentum is a highly mobile organ that moves with gut peristalsis in the small free volume of peritoneal fluid. Notably, the omentum has ability to occlude the sites of peritoneal inflammation, which prevent spreading of infection and trauma within the cavity (Meza-Perez and Randall 2017). Indeed, the omentum can efficiently collect bacteria from intestinal perforations (Ha et al. 2006), antigens from abdominal injuries (Ansel et al. 2002), and fluid from peritoneal dialysis (Beelen et al. 2005). Omentum is also the frequent site for metastasizing tumor cells such as ovarian cancer (Gerber et al. 2006). The ability of omentum to collect peritoneal antigens and cells is associated with migration of peritoneal macrophages, a phenomenon called macrophage disappearance reaction (MDR) (Barth et al. 1995). Despite macrophages within peritoneal cavity float within peritoneal fluid or mildly attach to parietal peritoneal wall in steady state, MDR is the physiological response against peritoneal irritation (e.g., bacterial infection) accompanied with rapid migration of macrophages that capture or engulf peritoneal antigens into the omentum (Okabe and Medzhitov 2014). Additionally, intraperitoneal challenge of IL-5, IL-10, TLR agonists or whole bacteria alters B-1 cell localization by modulating the patterns of adherence molecules including integrins and CD9, and induces their migration into the omentum (Carlow et al. 2009; Ha et al. 2006). Thus, omentum efficiently collects fluids, particles, bacteria, and cells from the peritoneal cavity which enable to induce subsequent immune responses (Morison 1906).

3 Omental Milky Spots

Although the omentum is primarily an adipose tissue, it contains a number of lymphoid aggregates (harbors up to 80 aggregates in homeostatic conditions of mouse) with 100–500 μm in diameter (Fig. 1a) (Benezech et al. 2015; Liu et al. 2016). These lymphoid aggregates, termed as "milky spots", which are first referred as white spots present in rabbit omenta by von Recklinghausen in 1863, and later characterized as the collections of immune cells, stromal cells, and structural elements surrounding glomerulus-like capillary beds (Meza-Perez and Randall 2017). Leukocyte populations of milky spots are mainly composed of macrophages and lymphocytes, which resemble to the composition of peritoneal cavity (Krist et al. 1995; Mebius 2009).

Accumulating evidences uncover a central role of omental milky spot in the regulation of adaptive immune responses in peritoneal cavity (Meza-Perez and Randall 2017). Omental B cells induce T cell dependent plasma cell responses upon intraperitoneal immunization (Dux et al. 1986; Hajdu et al. 1972), and the surgical removal of the omentum causes the reduced antibody responses to intraperitoneal immunization against sheep red blood cells (Rangel-Moreno et al. 2009). In addition, the evidences

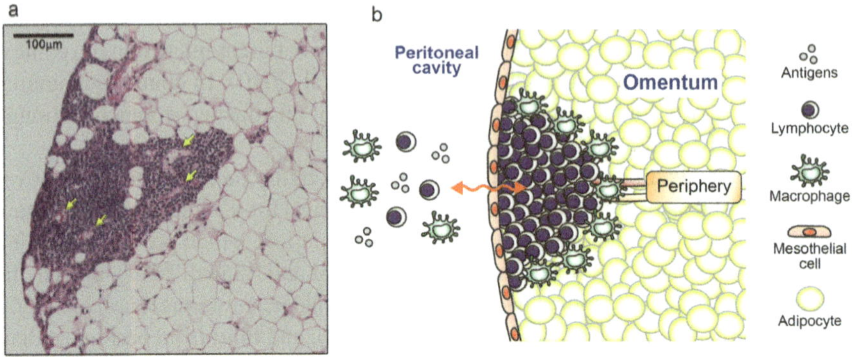

Fig. 1 **a** Omentum section was stained with hematoxylin and eosin (H&E). A milky spot is embedded between adipocytes and covered by a single layer of mesothelial cells (serous membrane). It is highly vascularized with blood vessels (arrows). Scale bar, 100 μm. **b** Schematic of milky spot. Macrophages are located around the milky spots. Peritoneal antigens and cells translocate between peritoneal cavity and milky spots through fenestrations (stomata) in the mesothelial layer

of T cell dependent B cell responses in the milky spots including isotype switching, somatic hypermutation, and affinity maturation have been demonstrated (Rangel-Moreno et al. 2009). Thus, the omental milky spot is the specialized place to regulate immune responses against peritoneal antigens, and it functions similar to secondary lymphoid organs.

4 Lymphatic Absorption of Peritoneal Fluid

In terms of the absorption of peritoneal fluid, lymphatic system of milky spots is quite distinct from that of the conventional secondary lymph nodes. Secondary lymph organs are covered by fibrous capsule that are thin and loose fibrous layer of collagenous material (Drayton et al. 2006). Numerous afferent lymph vessels that brand from the main vessels drain into a space underneath the capsule which is called subcapsular sinus. Subcapsular sinus is formed by CD169+ subcapsular sinus macrophages, lymphatic endothelial cells, and marginal reticular cells, and the roles of subcapsular sinus is to filter afferent lymph and display antigens from the incoming lymph to B cells (Moran et al. 2019).

In contrast, milky spots have neither capsule nor subcapsular sinus: they are only covered by single layer of serous membrane composed by mesothelial cells (Fig. 1b). Serous membrane which covers the surface of the milky spots is functionally and morphologically specialized to transport cells and antigens into lymphoid aggregates (Wang et al. 2010). Mesothelial cells that compose serous layer of the milky spots exhibit a cubic shape with small volume forming intercellular pores (5–20 μm in diameter), called "stomata", that are connected to lymphatic vessels drain peritoneal antigens and cells into milky spot (Mutsaers 2002; Yonemura and Yoshio Endou

2015). In contrast, mesothelial cells that cover the other area of omentum are flat, have larger volume with long microvilli, which provide tight connection of mesothelial layer.

In addition to the milky spots, the diaphragm composes another lymphatic drainage system specialized for peritoneal fluids, called "lymphatic lacunae" that are present between muscle fibers of the peritoneal side of the diaphragm (Li et al. 1996). Lymphatic capillaries of the lacunae are sophisticated to drain cells, antigens and peritoneal fluid with extremely flattened and broad lumina, connecting with opening (stomata) on the surface of diaphragm (Kim et al. 2009). In addition, transmural lymphatic branches connect the vessels of peritoneal and pleural sides (Abu-Hijleh et al. 1995). Therefore, the peritoneal fluid absorbed by lymphatic lacunae traverses the diaphragm to reach lymphatics beneath the diaphragmatic pleura. Altogether, lymphatic lacunae of diaphragm seem to compose main drainage channels for the lymphatic absorption from the peritoneal cavity in steady state, whereas omental milky spots collect peritoneal antigens and cells in response to peritoneal irritation and inflammation.

5 CXCL13 in Milky Spot Formation

Not only are milky spots lack of capsule and subcapsular sinus, but also the structure of lymphocyte follicle is also different between the milky spots and secondary lymph nodes. B cell follicles of milky spots do not appear to have centrally positioned networks of follicular dendritic cells (FDCs), despite the milky spots are capable to generate germinal center B cells that express BCL6 (Rangel-Moreno et al. 2009; Van Vugt et al. 1996). FDCs are non-haematopoietic stromal cells that reside in the center of B cell follicles, characterized as the expression of complement receptors CR1 (CD35) and CR2 (CD21) and the FDC marker FDCM1. FDCs do not express MHC (major histocompatibility complex) class II antigen molecules, but they are able to capture intact antigens on their cell surface for long periods, which is crucial for the selection of high-affinity antibody producing B cells during germinal center formation (Aguzzi et al. 2014). No apparent FDCs observed in B lymphocyte follicles may cause poor B cell selection and affinity maturation in milky spots (Rangel-Moreno et al. 2009).

FDCs in B cell follicles of lymph nodes secrete chemokine CXCL13 that attract B cells to build lymphoid network (Kranich and Krautler 2016). In contrast, the expression of CXCL13 in milky spots is uncharacteristically detected by several cell types including FDCM1 positive macrophages that locate outside of lymphoid clusters (Rangel-Moreno et al. 2009). CXCL13 is crucial in the recruitment of lymphocytes from bloodstream to the milky spots, which is illustrated by failure of milky spot formation in CXCL13 deficient mice (Ansel et al. 2002; Rangel-Moreno et al. 2009). High density of blood and lymphatic vessels runs down into the milky spots which contain PNAd-expressing high endothelial venules (HEVs) allow naïve lymphocytes to enter the milky spots from circulation (Rangel-Moreno et al. 2009). Although the precise cellular signaling mechanisms controlling lymphocyte migration is not clear, integrin $\alpha4\beta7$—mucosal addressin cell adhesion molecule 1 (MadCAM1) interaction enables conventional B cell (B-2 cells) recruitment from the blood into the milky spots (Berberich et al. 2008b).

Given that the milky spots recruit leukocytes from circulation, the milky spots are likely to act as a gateway for leukocytes entry into the peritoneal cavity (Ansel et al. 2002). In fact, omentectomy results in a 40% reduction of B cell migration from the circulation into the peritoneal cavity (Berberich et al. 2008a), and the egression of lymphocytes from omentum to the peritoneal cavity is mediated by sphingosin 1-phosphate (S1P) (Kunisawa et al. 2007). Integrin $\alpha4\beta1$ expressed on B cells is involved in the migration into the peritoneal cavity as well as their retention in the cavity (Berberich et al. 2008b). Taken together, milky spots mediate migration of the circulating leukocytes into the peritoneal cavity, in which chemokines, lipid mediators, and various cell adhesion molecules are involved (Cui et al. 2002).

6 Development of Milky Spots

Development of milky spots is hard-wired programmed which is taken place during embryogenesis: it starts to be appeared with the accumulation of monocytes and macrophages at 20 weeks of gestation in human, and true milky spots are detected at 35 weeks (Cruz-Migoni and Caamano 2016). The formation of milky spots follows distinct cues from that of secondary lymph nodes, which starts developing at 13 weeks of gestation. The initial steps of lymph node development involve clustering of lymphoid tissue inducer cells (LTi cells) that are hematopoietic cells expressing lymphotoxin-$\alpha\beta$ ($LT\alpha_1\beta_2$) with stromal organizer cells of mesenchymal origin (Roozendaal and Mebius 2011). $LT\alpha_1\beta_2$ from LTi cells can signal through Lymphotoxin-β receptor expressed on stromal organizer cells, leading to the induction of chemokines, cytokines, and adhesion molecules, which are required for the attraction and retention of additional hematopoietic cells and formation of lymph nodes. Due to the requirement of retinoic acid related orphan receptor (RORγt) and inhibitor of differentiation-2 (Id2) for LTi differentiation, mice deficient in RORγt or Id2 genes lack secondary lymphoid organs. In contrast, well-development of milky spots in mice deficient in either of RORγt and Id2 genes shows that LTi cells are dispensable for milky spot formation (Rangel-Moreno et al. 2009). Although the formation of milky spots is strongly diminished in LTα deficient mice, this is not caused by the developmental defect since the milky spot formation can be completely restored by the reconstitution of LTα sufficient hematopoietic cells in adult mice (Rangel-Moreno et al. 2009). Thus, the milky spots do not require interaction with LTi cells, despite what type of cells replace functions of LTi cells remains to be determined.

Of note, the size and number of milky spots can increase by the irritations of peritoneal cavity such as infection and peritoneal dialysis (Benezech et al. 2015; Rangel-Moreno et al. 2009). This suggests that milky spots share the characteristics with tertiary lymphoid organs (TLOs), the accumulation of lymphoid and stromal cells that arise and organize at ectopic sites upon chronic inflammation such as autoimmunity, infection, graft rejection, and cancer (Ruddle 2014). Thus, milky spots share characteristics of both developmentary programmed-secondary lymphoid organs and ectopically induced-tertiary lymphoid organs (Meza-Perez and Randall 2017).

7 Fat-Associated Lymphoid Clusters

Recently, lymphoid clusters that are reminiscent of omental milky spots have been identified in mesenteric fat as well as gonadal, mediastinal, and pericardial fats (Cruz-Migoni and Caamano 2016; Kaminski and Randall 2010). These structures, called fat-associated lymphoid clusters (FALCs), are originally discovered in mesentery as the lymphoid clusters containing a population (20–40% of FALC cells) of group 2 innate lymphoid cells (ILC2s) characterized as lineage markers-negative but progenitor potential markers positive cells (Moro et al. 2010). ILC2s can secrete multiple type 2 (Th2) associated cytokines, including IL-4, -5, -6, and -13, that are important for the protection from parasite infection. FALCs share similar characteristics with milky spots in terms of their structures: they are not encapsulated by a fibrous capsule and an underlying subcapsular sinus, the lymphoid clusters are in close interaction with surrounded adipocytes (Moro et al. 2010). Similar to milky spots, FALC formation is independent on LTi cells as shown by the occurrence of FALCs in RORγt deficient mice (Benezech et al. 2015). On the other hands, development of FALCs in mice is postnatally, which is contrary to the formation of the milky spots that develop during embryogenesis. The formation of FALCs is identified at around 2–3 weeks after birth, increasing their numbers to reach a plateau at around 18 weeks of age (Benezech et al. 2015). Although FALC stromal cells that produce high amounts of CXCL13, CXLC13 are not important for FALC formation since FALCs in mice deficient in CXCR5, the receptor for CXCL13, are well-developed although the number of B lymphocyte present in the clusters is markedly reduced (Benezech et al. 2015). Overall, the milky spots and FALCs are structurally similar, but have distinct requirement for instructive signals for their development.

8 Milky Spots Provide Signals for Macrophage Differentiation/Polarization

Surgical removal of omentum in adult rats results in the reduction in the number of peritoneal macrophages (Agalar et al. 1997), and clodronate liposome mediated depletion of macrophages induce local proliferation of macrophages around milky spots (Ratajczak et al. 1987). These evidences suggest milky spots play central roles in the differentiation and maintenance of peritoneal macrophages. Indeed, stromal cells of milky spots are rich sources of macrophage colony stimulating factor (M-CSF), which is crucial for differentiation and maintenance of macrophages. Additionally, transcription factor Wilms' Tumor 1 (WT1)—expressing mesothelial and fibroblastic stromal cells in omentum—produces retinoic acid which induce zinc finger transcription factor GATA6 (Buechler et al. 2019). GATA6 acts as master transcriptional regulator for functional specialization, controlling peritoneum specific gene expression program which is essential for immune regulation, cell adhesion, and local proliferation/self-renewal of peritoneal macrophages (Gautier et al. 2014; Okabe and Medzhitov 2014; Rosas et al. 2014). GATA6 also regulate macrophage migration from omentum to the peritoneal cavity (Okabe and Medzhitov 2014). These evidences suggest milky spots assign specific anatomical location where M-CSF and retinoic acid accumulate, essential for differentiation, proliferation and functional polarization of peritoneal macrophages (Okabe and Medzhitov 2016).

In this regard, WT1 expressing stromal cells (fibroblasts and mesothelial cells) that produce retinoic acid (and M-CSF, presumably) could be denoted as "niche cells" as a key source of factors instructing macrophage phenotypes (Fig. 2). Interestingly, retinoic acid is required for the induction of CXCL13 in peritoneal macrophages (Okabe and Medzhitov 2014), reminiscent with FDCs which require retinoic acid for the induction of CXCL13 (Koning et al. 2020). Interestingly, omental and peritoneal macrophages harbor several characteristics of FDCs in addition to the production of CXCL13 which is essential for B lymphocyte recruitment and lymphoid follicle formation. Peritoneal macrophages express FDC marker FDCM1 (alternatively, milk fat globule epidermal growth factor 8; MFG-E8) which is tethering

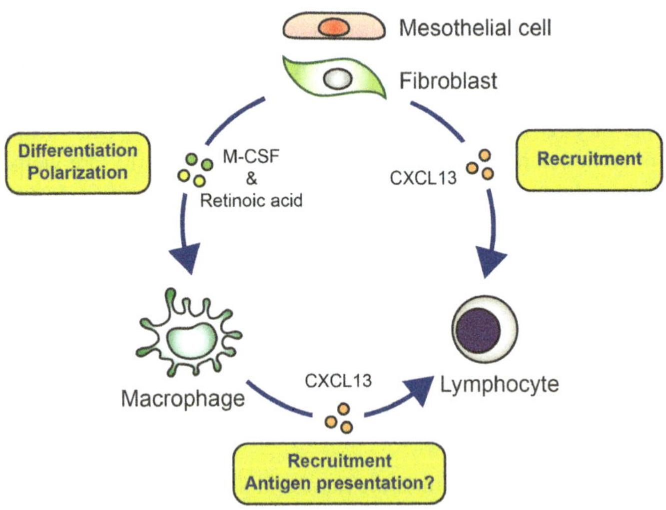

Fig. 2 Schematic overview of the cellular network in milky spot. Speculated cell–cell network with the factors that mediate the communication is indicated, based on the literature. This overview is simplified for clarity and the involvement of additional cell types and factors remain to be verified experimentally

molecule bridge between integrins and phosphatidylserine (Kranich et al. 2008). Furthermore, GATA6-expressing peritoneal macrophages suppress their phagocytic activity (Irvine et al. 2016) as well as they have negligible expression of MHC class II antigen. Given that B-1 lymphocytes can produce antibodies without T cell help, these macrophages, similar to FDCs, may have roles in carrying peritoneal antigens that are held on their cell surface and expose to B-1 cells for activation.

9 Conclusion

Peritoneal cavity composes unique immune subsets that are not only important for the protection of body cavity, but it is also associated with visceral tissue homeostasis and peripheral immune control. In this review, I have highlighted the roles of omental milky spot in the development and regulation of peritoneal immunity. Omental milky spots have central roles in the regulation of peritoneal immune responses with absorption of peritoneal antigens, recruitment of leukocytes, controlling T cell dependent/independent antibody production, and regulation of differentiation and polarization of macrophages. The precise cellular mechanisms controlling peritoneal immunity as well as the influence of the perturbation of the omentun functions such as peritoneal dialysis, tumor metastasis, and omentectomy may facilitate the design of therapeutics targeting peritoneal immune systems.

References

Abu-Hijleh MF, Habbal OA, Moqattash ST (1995) The role of the diaphragm in lymphatic absorption from the peritoneal cavity. J Anat 186(Pt 3):453–467

Agalar F, Sayek I, Cakmakci M, Hascelik G, Abbasoglu O (1997) Effect of omentectomy on peritoneal defence mechanisms in rats. Eur J Surg 163:605–609

Aguzzi A, Kranich J, Krautler NJ (2014) Follicular dendritic cells: origin, phenotype, and function in health and disease. Trends Immunol 35:105–113

Ansel KM, Harris RB, Cyster JG (2002) CXCL13 is required for B1 cell homing, natural antibody production, and body cavity immunity. Immunity 16:67–76

Barth MW, Hendrzak JA, Melnicoff MJ, Morahan PS (1995) Review of the macrophage disappearance reaction. J Leukoc Biol 57:361–367

Baumgarth N (2011) The double life of a B-1 cell: self-reactivity selects for protective effector functions. Nat Rev Immunol 11:34–46

Baumgarth N (2016) B-1 cell heterogeneity and the regulation of natural and antigen-induced IgM production. Front Immunol 7:324

Beelen RH, Oosterling SJ, van Egmond M, van den Born J, Zareie M (2005) Omental milky spots in peritoneal pathophysiology (spots before your eyes). Perit Dial Int 25:30–32

Benezech C, Luu NT, Walker JA, Kruglov AA, Loo Y, Nakamura K, Zhang Y, Nayar S, Jones LH, Flores-Langarica A et al (2015) Inflammation-induced formation of fat-associated lymphoid clusters. Nat Immunol 16:819–828

Berberich M, Krause AM, Orlandi M, Scandola F, Wurthner F (2008a) Toward fluorescent memories with nondestructive readout: photoswitching of fluorescence by intramolecular electron transfer in a diaryl ethene-perylene bisimide photochromic system. Angew Chem Int Ed Engl 47:6616–6619
Berberich S, Dahne S, Schippers A, Peters T, Muller W, Kremmer E, Forster R, Pabst O (2008b) Differential molecular and anatomical basis for B cell migration into the peritoneal cavity and omental milky spots. J Immunol 180:2196–2203
Buechler MB, Kim KW, Onufer EJ, Williams JW, Little CC, Dominguez CX, Li Q, Sandoval W, Cooper JE, Harris CA et al (2019) A stromal niche defined by expression of the transcription factor WT1 mediates programming and homeostasis of cavity-resident macrophages. Immunity 51:119–130 e115
Carlow DA, Gold MR, Ziltener HJ (2009) Lymphocytes in the peritoneum home to the omentum and are activated by resident dendritic cells. J Immunol 183:1155–1165
Cruz-Migoni S, Caamano J (2016) Fat-associated lymphoid clusters in inflammation and immunity. Front Immunol 7:612
Cui L, Johkura K, Liang Y, Teng R, Ogiwara N, Okouchi Y, Asanuma K, Sasaki K (2002) Biodefense function of omental milky spots through cell adhesion molecules and leukocyte proliferation. Cell Tissue Res 310:321–330
Di Nicola V (2019) Omentum a powerful biological source in regenerative surgery. Regen Ther 11:182–191
Drayton DL, Liao S, Mounzer RH, Ruddle NH (2006) Lymphoid organ development: from ontogeny to neogenesis. Nat Immunol 7:344–353
Dux K, Rouse RV, Kyewski B (1986) Composition of the lymphoid cell populations from omental milky spots during the immune response in C57BL/Ka mice. Eur J Immunol 16:1029–1032
Gautier EL, Ivanov S, Williams JW, Huang SC, Marcelin G, Fairfax K, Wang PL, Francis JS, Leone P, Wilson DB et al (2014) Gata6 regulates aspartoacylase expression in resident peritoneal macrophages and controls their survival. J Exp Med 211:1525–1531
Gerber SA, Rybalko VY, Bigelow CE, Lugade AA, Foster TH, Frelinger JG, Lord EM (2006) Preferential attachment of peritoneal tumor metastases to omental immune aggregates and possible role of a unique vascular microenvironment in metastatic survival and growth. Am J Pathol 169:1739–1752
Ha SA, Tsuji M, Suzuki K, Meek B, Yasuda N, Kaisho T, Fagarasan S (2006) Regulation of B1 cell migration by signals through toll-like receptors. J Exp Med 203:2541–2550
Hajdu I, Holub M, Trebichavsky I (1972) The sequence of appearance of antibodies in mouse omentum plasma cells. Exp Cell Res 75:219–230
Healy JC, Reznek RH (1998) The peritoneum, mesenteries and omenta: normal anatomy and pathological processes. Eur Radiol 8:886–900
Irvine KM, Banh X, Gadd VL, Wojcik KK, Ariffin JK, Jose S, Lukowski S, Baillie GJ, Sweet MJ, Powell EE (2016) CRIg-expressing peritoneal macrophages are associated with disease severity in patients with cirrhosis and ascites. JCI Insight 1, e86914
Isaza-Restrepo A, Martin-Saavedra JS, Velez-Leal JL, Vargas-Barato F, Riveros-Duenas R (2018) The peritoneum: beyond the tissue: a review. Front Physiol 9:738
Kaminski DA, Randall TD (2010) Adaptive immunity and adipose tissue biology. Trends Immunol 31:384–390
Kim KE, Koh YJ, Jeon BH, Jang C, Han J, Kataru RP, Schwendener RA, Kim JM, Koh GY (2009) Role of CD11b+ macrophages in intraperitoneal lipopolysaccharide-induced aberrant lymphangiogenesis and lymphatic function in the diaphragm. Am J Pathol 175:1733–1745
Koning JJ, Rajaraman A, Reijmers RM, Konijn T, Pan J, Ware CF, Butcher EC, Mebius RE (2020) Development of follicular dendritic cells in lymph nodes depends on retinoic acid mediated signaling. bioRxiv
Kranich J, Krautler NJ (2016) How follicular dendritic cells shape the B-cell antigenome. Front Immunol 7:225

Kranich J, Krautler NJ, Heinen E, Polymenidou M, Bridel C, Schildknecht A, Huber C, Kosco-Vilbois MH, Zinkernagel R, Miele G et al (2008) Follicular dendritic cells control engulfment of apoptotic bodies by secreting Mfge8. J Exp Med 205:1293–1302

Krist LF, Eestermans IL, Steenbergen JJ, Hoefsmit EC, Cuesta MA, Meyer S, Beelen RH (1995) Cellular composition of milky spots in the human greater omentum: an immunochemical and ultrastructural study. Anat Rec 241:163–174

Kunisawa J, Kurashima Y, Gohda M, Higuchi M, Ishikawa I, Miura F, Ogahara I, Kiyono H (2007) Sphingosine 1-phosphate regulates peritoneal B-cell trafficking for subsequent intestinal IgA production. Blood 109:3749–3756

Li J, Zhao Z, Zhou J, Yu S (1996) A study of the three-dimensional organization of the human diaphragmatic lymphatic lacunae and lymphatic drainage units. Ann Anat 178:537–544

Liu J, Geng X, Li Y (2016) Milky spots: omental functional units and hotbeds for peritoneal cancer metastasis. Tumour Biol 37:5715–5726

Mebius RE (2009) Lymphoid organs for peritoneal cavity immune response: milky spots. Immunity 30:670–672

Meza-Perez S, Randall TD (2017) Immunological functions of the omentum. Trends Immunol 38:526–536

Moran I, Grootveld AK, Nguyen A, Phan TG (2019) Subcapsular sinus macrophages: the seat of innate and adaptive memory in murine lymph nodes. Trends Immunol 40:35–48

Morison R (1906) Remarks on some functions of the omentum. Br Med J 1:76–78

Moro K, Yamada T, Tanabe M, Takeuchi T, Ikawa T, Kawamoto H, Furusawa J, Ohtani M, Fujii H, Koyasu S (2010) Innate production of T(H)2 cytokines by adipose tissue-associated c-Kit(+)Sca-1(+) lymphoid cells. Nature 463:540–544

Mutsaers SE (2002) Mesothelial cells: their structure, function and role in serosal repair. Respirology 7:171–191

O'Connell J, Lynch L, Cawood TJ, Kwasnik A, Nolan N, Geoghegan J, McCormick A, O'Farrelly C, O'Shea D (2010) The relationship of omental and subcutaneous adipocyte size to metabolic disease in severe obesity. PLoS One 5:e9997

Okabe Y, Medzhitov R (2014) Tissue-specific signals control reversible program of localization and functional polarization of macrophages. Cell 157:832–844

Okabe Y, Medzhitov R (2016) Tissue biology perspective on macrophages. Nat Immunol 17:9–17

Rangel-Moreno J, Moyron-Quiroz JE, Carragher DM, Kusser K, Hartson L, Moquin A, Randall TD (2009) Omental milky spots develop in the absence of lymphoid tissue-inducer cells and support B and T cell responses to peritoneal antigens. Immunity 30:731–743

Ratajczak MZ, Jaskulski D, Pojda Z, Wiktor-Jedrzejczak W (1987) Omental lymphoid organ as a source of macrophage colony stimulating activity in peritoneal cavity. Clin Exp Immunol 69:198–203

Roberts AW, Lee BL, Deguine J, John S, Shlomchik MJ, Barton GM (2017) Tissue-resident macrophages are locally programmed for silent clearance of apoptotic cells. Immunity 47:913–927 e916

Roozendaal R, Mebius RE (2011) Stromal cell-immune cell interactions. Annu Rev Immunol 29:23–43

Rosas M, Davies LC, Giles PJ, Liao CT, Kharfan B, Stone TC, O'Donnell VB, Fraser DJ, Jones SA, Taylor PR (2014) The transcription factor Gata6 links tissue macrophage phenotype and proliferative renewal. Science 344:645–648

Ruddle NH (2014) Lymphatic vessels and tertiary lymphoid organs. J Clin Invest 124:953–959

Szender JB, Emmons T, Belliotti S, Dickson D, Khan A, Morrell K, Khan A, Singel KL, Mayor PC, Moysich KB et al (2017) Impact of ascites volume on clinical outcomes in ovarian cancer: a cohort study. Gynecol Oncol 146:491–497

Van Vugt E, Van Rijthoven EA, Kamperdijk EW, Beelen RH (1996) Omental milky spots in the local immune response in the peritoneal cavity of rats. Anat Rec 244:235–245

Wang J, Kubes P (2016) A reservoir of mature cavity macrophages that can rapidly invade visceral organs to affect tissue repair. Cell 165:668–678

Wang ZB, Li M, Li JC (2010) Recent advances in the research of lymphatic stomata. Anat Rec (hoboken) 293:754–761

Wilkosz S, Ireland G, Khwaja N, Walker M, Butt R, de Giorgio-Miller A, Herrick SE (2005) A comparative study of the structure of human and murine greater omentum. Anat Embryol (berl) 209:251–261

Yonemura Y, Yoshio Endou EC (2015) Mechanisms of the formation of peritoneal surface malignancy on omental milky spots from low grade appendiceal mucinous carcinoma. J Clin Exp Oncol 03

Yoo E, Kim JH, Kim MJ, Yu JS, Chung JJ, Yoo HS, Kim KW (2007) Greater and lesser omenta: normal anatomy and pathologic processes. Radiographics 27:707–720

The manufacturer's authorised representative in the EU is Springer Nature Customer Service Centre GmbH, Europaplatz 3, 69115 Heidelberg, Germany. If you have any concerns regarding our products, please contact ProductSafety@springernature.com

Printed and bound by CPI Group (UK) Ltd, Croydon, CR0 4YY
28/09/2026
02224074-0016